Alexander D. Anderson

The Tehuantepec Inter-Ocean Railroad

A commercial and statistical review showing its local, national, and

international features, and advantages

Alexander D. Anderson

The Tehuantepec Inter-Ocean Railroad
A commercial and statistical review showing its local, national, and international features, and advantages

ISBN/EAN: 9783337317881

Printed in Europe, USA, Canada, Australia, Japan

Cover: Foto ©berggeist007 / pixelio.de

More available books at **www.hansebooks.com**

THE TEHUANTEPEC

INTER-OCEAN RAILROAD,

A COMMERCIAL AND STATISTICAL REVIEW SHOWING ITS
LOCAL, NATIONAL, AND INTERNATIONAL FEATURES,
AND ADVANTAGES.

BY

ALEX. D. ANDERSON,

AUTHOR OF "THE SILVER COUNTRY, OR THE GREAT SOUTHWEST;" AND "THE
MISSISSIPPI AND TRIBUTARIES."

A. S. BARNES & COMPANY,

NEW YORK AND CHICAGO.

1880.

TO

HON. EDWARD LEARNED

OF MASSACHUSETTS,

Whose energetic efforts insure the immediate consummation of Inter-Ocean Transit across the Isthmus of Tehuantepec—an enterprise which will supplement the Mississippi River and practically extend it to the Pacific Ocean, give to the United States an Isthmus Route pre-eminently the most advantageous to her commerce, inaugurate in Mexico a new era of commercial progress, stimulate closer intercourse between the two sister Republics, and which merits the co-operation of all interested in the material advancement of America—these pages are respectfully inscribed.

PREFACE.

IT is not the purpose of this paper to engage in a discussion of the various schemes for a ship-canal across the American Isthmus, but to consider the merits of the route which is nearest our shores, most easily protected, and preeminently the most serviceable to American Commerce.

An Inter-Oceanic Railway is already under construction across the Isthmus of Tehuantepec, and will soon be ready to transport from sea to sea the interchanges between the two coasts of the New World, and the older nations on the East and West. It is to the very superior advantages of this route that the attention of the public is invited.

The following pages have been prepared in legal brief style with a citation of authorities ; first, because the value of any statistics depends largely upon knowing the sources of their derivation ; second, that the reader who dwells on a less favored part of the earth may know that the statements concerning the marvelous fertility of the Isthmus are collected from official and other standard sources, and not works of

fiction ; and finally, because this is simply a summary of leading facts, and the reader may wish to examine more in detail
some of the many advantages of this commercial highway. If
so, the authorities cited in the foot-notes and in the appendix
will furnish a great variety of information.

WASHINGTON, D. C., *Oct.* 28, 1880.

CONTENTS.

CHAPTER I.

PAGE

HISTORICAL NOTES................. 9

MAP OF THE ISTHMUS 19

CHAPTER II.

LOCAL FEATURES.

Descriptive Notes... 19

Climate and Health... 21

Products 23

Coffee 24		Timber................... 30
Sugar.. 25		India-rubber 32
Oranges 26		Vegetable Dyes 33
Corn 27		Indigo................... 33
Bananas 27		Cochineal................ 34
Cotton................... 28		Petroleum 34
Tobacco,................. 29		Ixtle or Pita 35
Rice..................... 30		Cattle 36
Cocoa.................... 30		Fish and Game.......... 37

Water Power and Manufactures 38

Population.. 39

Antiquities.. 39

Harbors and Railway Line 43

Land Grant.. 46

MAP ILLUSTRATING RELATIONS TO MEXICO 47

CHAPTER III.

COMMERCIAL RELATIONS TO MEXICO.

Need of Railways 47

Topographical Features.. 49

A Railway Era... 50

The Isthmus as an Outlet...... 53

As a Transcontinental Highway 56

PAGE

MAP ILLUSTRATING RELATIONS TO THE UNITED STATES 58

CHAPTER IV.

COMMERCIAL RELATIONS TO THE UNITED STATES.

To the Mississippi Valley 58
To the Valley and Gulf States combined.......................... 62
To the Atlantic States 63
To the Pacific States... 65

MAP ILLUSTRATING RELATIONS TO THE WORLD 67

CHAPTER V.

COMMERCIAL RELATIONS TO THE WORLD.

Its Central Position...................... 67
Possible Present Patronage....................................... 69
Possible Future Patronage 69
    From the Natural Growth of Commerce....................... 71
    From Exceptional Causes 73

CHAPTER VI.

POLITICAL FEATURES.............. 77

APPENDIX.

Table of Distances..... 81
List of Authorities.. 85
    Surveys 85
    Popular Publications 86
    Congressional Documents 88

CHAPTER I.

THE importance of the Isthmus of Tehuantepec may readily be seen from a glance at its varied history during a period of more than three hundred and sixty years, under Indian, Spanish, and Mexican rule.

In 1520, as soon as Cortez had, through the hospitality of Montezuma, become installed in the Aztec national palace, and before the bloody conflict which resulted in the conquest and destruction of the ancient city of Mexico, he sent out exploring parties to discover the mines of precious metals, and to find on the Gulf coast a better harbor. In his dispatches to his king, Charles V of Spain, Cortez tells the story as follows : "I likewise inquired of Muteczuma if there were, on the coast of the sea, any river, or bay, into which ships could enter and lie with safety. He answered that he did not know, but that he would cause a chart of the coast to be painted showing the rivers and bays, and that I might send Spaniards to examine them, for which purpose he would dispatch suitable persons with them as guides ; and he did so. The next day they brought me a chart of the whole coast, painted on cloth, on which appeared a river that discharged into the sea with a wider mouth, according to the chart, than any others." [1]

He further relates that he sent "ten men, and among them several pilots, and persons acquainted with the sea," [2] to examine this river Coatzacoalcos—that they departed from Vera Cruz, and proceeded along the coast until they reached its

[1] The Dispatches of Hernando Cortes, p. 99.   [2] Ditto, p. 100.

mouth—that "they found two fathoms and a half of water at its entrance, in the shallowest part, and ascending twelve leagues, the least depth they found was five or six fathoms ;" [1] also that they reported "there were numerous and large towns on its banks, and the whole province was level and well fortified, rich in all the productions of the earth, and containing a numerous population who are not vassals, or subjects of Muteczuma, but rather his enemies." [2]

Cortez then adds the following comment : " When I was informed by the Spaniards that the province they had visited was in a situation to be colonized, and that they had discovered a harbor in it, I was much gratified, since from the time that I had first set foot in this country I had constantly sought to find some harbor upon its coast where I might found a settlement, but I had not been able to discover one, nor is there any on the coast from the river San Antonio, which is next to Grijalva, to that of Panuco, which is down the coast." [3] The river Grijalva is now called Tabasco, at its mouth, and is in the State of Tabasco. The River Panuco is in the State of Tamaulipas, and empties into the harbor of Tampico.

Immediately upon the return of this exploring party he sent out another expedition of one hundred and fifty men, "for the purpose of tracing, planning, and settling the town, and erecting a fortress." [4] Soon after he sent out, for the same purpose, another expedition of four hundred men. [5] Later, probably in the year 1523, he sent still another expedition to the Isthmus under the lead of Sandoval. [6] One of the party was Bernal Diaz, who afterward became the Spanish historian of the conquest. He, with several others of the expedition, settled at Guacasualco, and remained there until Cortez himself passed through that place, in October 1524, when they joined his party on an excursion still farther south. [7]

---

[1] The Dispatches of Hernando Cortes, p. 100.   [2] Ditto, p. 100.
[3] Ditto, p. 101.   [4] Ditto, p. 102.   [5] Ditto, p. 144.
[6] The Memoirs of the Conquistador Bernal Diaz, vol. ii, p. 141.
[7] Prescott's Conquest of Mexico, vol. iii, p. 265.

The Conqueror improved this opportunity to visit the birth-place of his celebrated mistress, Doña Marina, called by the Aztecs Malinche.[1] She was born at the village of Jaltipan (formerly known as Painala), in the northern division of the Isthmus.[2] In return for her invaluable services to the Spaniards, as interpretess, and secretary, in negotiations with the native officials, and otherwise, she received large grants of land, part of which were located on the Isthmus, near her former home.[3]

Cortez's chief ambition at this time seems to have been the discovery of a highway between the two oceans, for, in his fourth letter to Charles V, dated October, 15, 1524, he says, "Being well aware of the great desire of your majesty to know concerning the supposed strait, and of the great advantage the crown would derive from its discovery, I have laid aside all other schemes, more obviously tending to promote my interests, in order to pursue this object alone."[4]

A few years later, in 1529, he was made Marquis of the Valley of Oaxaca, in the state of the same name, which the Isthmus of Tehuantepec intersects. He also received from the king a grant of large estates.[5] Part of these lands, situated in the central division of the Isthmus, are alluded to by J. McL. Murphy, of the U. S. Navy, who visited the Isthmus, in 1851, with the Scientific Commission sent there to make a railway survey. He said of these estates, in a lecture before the American Geographical Society : "From these gigantic hillocks " (Maletengo Hills) "were spread out before us, like a map, the *estates of Marquesanas*—the gift lands of Charles V to the Conqueror of Mexico—stretching away to the south, even beyond the sharp, angular peaks of the dividing ridge. All through this vast estate, which embraces an area of somewhat less than 200,000

---

[1] Diaz's Memoirs, vol. i, p. 84.
[2] Ditto, p. 84 ; and "The Isthmus of Tehuantepec," by J. McL. Murphy, in Journal of American Geographical and Statistical Society for 1859, p. 162.
[3] Prescott's Conquest, vol. iii, p. 279.
[4] Cortez's Dispatches, p. 419.
[5] Prescott's Conquest of Mexico, vol. iii, pp. 306–7.

acres, there are ruins of indigo-vats, lime-kilns, corrals, and other improvements."[1] Another part of his marquisate comprised the town of Tehuantepec, on the Pacific side of the Isthmus, and at that place, together with the harbors of Zacatula and Acapulco, was fitted out his expedition on the Pacific in 1533, "according to his agreement with the Empress Isabella, of glorious memory, and with the Council of the Indies, for discoveries in the South Sea."[2]

Before sailing he sought to provide a commercial highway across the Isthmus. Says a Mexican authority: "It was he who conceived the idea of a lucrative speculation by means of a road over the Isthmus to supply Spain with the spices of the East Indies, and the products of such new regions as he expected to discover."[3] It is evident that he constructed a military road from the town of Tehuantepec across to the head of navigation of the Coatzacoalcos River, and that it remained in use for a period of about one hundred years, for these facts are stated in a carefully prepared lecture before the American Geographical Society by one who explored the Isthmus as a member of Barnard and Williams' surveying party.[4]

A century later, citizens of Oaxaca presented a petition to the Viceroy, under Spain, of Mexico, asking that the harbor of Coatzacoalcos might be made "a port of entry, and a great depot of commerce instead of the port and city of Vera Cruz."[5] In it they also suggest that a canal or road be constructed across the Isthmus.[6]

In 1774 Don Augustin Cramer, a civil engineer, made, by order of the Viceroy of Mexico, a voyage of discovery. In his report he said of Tehuantepec: "It would not be a work of great difficulty, nor excessively costly, to effect a communication

[1] Journal of Amer. Geog. and Statistical Soc. for 1859, p. 174.
[2] Diaz's Memoirs, vol. ii, p. 349.
[3] The Isthmus of Tehuantepec, by Garay, pp. 27-8.
[4] Jour. of Amer. Geog. Soc., June 1859, p. 164.
[5] Memoirs of the Mexican Revolution, by W. D. Robinson, pp. 359-60.
[6] Ditto, p. 360.

between the two seas across this Isthmus."[1] Don José de Garay, from whose pamphlet we have just quoted, says of the next movement in favor of a canal, in the year 1814 : "The Spanish Cortes, having in view all that had been written upon this subject, and the various reports presented by persons charged with the surveys of the principal points suggested for the communication, by a decree of the 30th of April, 1814, authorized the opening of a canal across the Isthmus of Tehuantepec in preference to those of Nicaragua and Panama."[2]

In 1824, three years after Mexico had become an independent republic, the state of Vera Cruz, and the general government of Mexico, appointed respectively Don Tadeo de Ortiz and Col. Don Juan de Orbegozo to survey this Isthmus. Each party made a report.[3]

On the 1st of March, 1842, the President of Mexico, General Santa Anna, granted to Don José de Garay, a citizen of that republic, the right of way across Tehuantepec for communication between the Atlantic and Pacific. Article 2 of the grant says : "This shall be performed by water ; but when this may not be convenient, then railroads and steam carriages may be used."[4] At the commencement of the decree Gen. Santa Anna uses the following significant language : " Know ye that, firm to my purpose of exalting the nation, and of rendering the people happy, and taking into consideration the proposition which Don José Garay has presented, and considering that no means are so sure and effectual for promoting the national prosperity as that of creating the republic the centre of communication and navigation of all countries,"[5] etc. On the 30th of April following, Gaetano Moro, the engineer appointed by the grantee, started on the surveying expedition.

In 1844 Mexico extended the time for beginning the construction for one year, ending July 1st, 1845.[6]

On the 5th of November, 1846, the acting President of Mexico extended the grant to Garay for two years from that date.[7]

[1] An account of the Isthmus of Tehuantepec, by Don José de Garay, pp. 28-9. [2] Ditto, p. 23. [3] Ditto, p. 30. [4] Ditto, p. 105. [5] Ditto, p. 104. [6] Senate Ex. Doc. 72, Thirty-fifth Cong. First Sess., p. 40. [7] Ditto, p. 41.

On the 15th of April, 1847, during the negotiations between the United States and Mexico, which finally resulted in the Treaty of Guadalupe Hidalgo of 1848, James Buchanan, then Secretary of State, sent the following instructions to Nicholas P. Trist, the commissioner of the United States at the City of Mexico, viz.: "Instead of fifteen millions of dollars stipulated to be paid by the fifth article for the extension of our boundary over New Mexico and Upper and Lower California, you may increase the amount to any sum not exceeding thirty millions of dollars, payable by instalments of three millions per annum, provided the right of passage and transit across the Isthmus of Tehuantepec, secured to the United States by the eighth article of the projét, shall form a part of the treaty." [1]

This offer shows how important the United States then considered inter-oceanic communication at the Isthmus, but it was rejected by Mexico for the sufficient reason that she had already granted the right of way to one of her own citizens, and was not, therefore, at liberty to grant the same to the United States.[2]

On the 28th of September, 1848, Don José de Garay transferred his grant to Messrs. Manning and McIntosh, citizens of Great Britain.[3]

On the 5th of February, 1849, the grant was transferred by them to a citizen of the United States, one Peter A. Hargous.[4]

In December, 1850, the Scientific Commission under the command of Major J. G. Barnard of the U. S. Engineers, assisted by J. J. Williams, C. E., landed at the Isthmus and immediately commenced a thorough survey for the purpose of an inter-oceanic railway. This was in behalf of the Tehuantepec Railroad Company of New Orleans, organized under the assignment to Hargous.

On the 23d of May, 1851, the Mexican Congress declared

---

[1] Senate Ex. Doc. No. 60, Thirtieth Congress, First Session, p. 44.
[3] Senate Ex. Doc. No. 72, Thirty-fifth Congress, First Session, p. 12.
[2] Senate Report 355, Thirty-second Congress, First Session.
[4] Ditto.

the extension of the grant issued by the Acting President on the 5th of November, 1846, to be null and void.[1]

On the 14th of May, 1852, the Mexican Congress provided for an invitation of proposals for opening a highway across the Isthmus.[2]

Under this Act, on the 5th of February, 1853, the President of Mexico awarded the contract to a Mixed Company—part Mexicans, and part citizens of the United States—of which, one A. G. Sloo was the leading spirit.[3]

On the 30th of December, 1853, a treaty (commonly known as the Gadsden Treaty) was concluded between the United States and Mexico, which, in Article VIII, after referring to the Sloo grant just mentioned, guaranteed the right of transit to the United States and its citizens.[4]

In 1855, November 26th, the President of Mexico recognized one F. de P. Falconet, a British subject, as the sole owner of the grant to the Mixed Company, it having been pledged to him as security for the repayment of $600,000 which he had advanced to Mexico in behalf of the Company.[5]

On the 3d of September, 1857, the President of Mexico annulled the contract with Sloo and Associates, composing the Mixed Company.[6]

On the 7th of September, 1857, President Comonfort granted the right of way for a railroad to the Louisiana Tehuantepec Company.[7]

On the 28th of March, 1859, Mexico extended the time of the grant last mentioned.[8]

On the 25th of October, 1860, this grant was again extended by Mexico.[9]

In 1861 Napoleon III instructed his Minister in Mexico to secure the grant which the Louisiana Company had forfeited.[10]

[1] Senate Ex. Doc. No 72, Thirty-fifth Congress, First Session, p. 41.
[2] Ditto, p. 41.   [3] Ditto, p. 41.   [4] U. S. Statutes, vol. x, pp. 1031–37.
[5] Sen. Ex. Doc. No. 72, Thirty-fifth Cong., 1st Sess., pp. 41–2.   [6] Do., p. 54.
[7] Ditto, p. 55.   [8] Sen. Ex. Doc. 25, Thirty-ninth Cong., 2d Sess., p. 10.
[9] Ditto, p. 10.   [10] Commodore Shufeldt, in N. Y. *Herald*, Oct. 22, 1879.

On the 12th of October, 1866, the Emperor of Mexico, Maximilian, extended the time of the grant formerly made by the Republic of Mexico to the Louisiana Company.[1]

Three days later, on October 15th, President Juarez, not recognizing the authority of Maximilian, then in possession of the capital, granted the right of way for a railroad and telegraph line to the Tehuantepec Transit Company. By the same instrument he annulled the grant of September 7th, 1857, and extensions of the same, because of an infringement of its obligations. This new Company had its head-quarters at New York, and Charles Knap was its President.[2] This grant was subsequently declared to be forfeited,[3] and

On the 6th of December, 1867, a grant was made by the President of Mexico, to Emilio de la Sere, a citizen of the United States, of the right of way across the Isthmus.[4]

On the 10th of November, 1868, the General Assembly of Vermont passed an Act establishing the incorporation of the Tehuantepec Railway Company organized on the strength of the last mentioned concession from Mexico.[5]

On the 29th of December, 1868, this grant was modified by Act of the Mexican Congress.[6]

On the 18th of February, 1869, La Sere conveyed his grant to said Tehuantepec Railway Company.

On the 14th of December, 1870, this Company was, by Act of the Mexican Congress, also granted the right to construct a canal across the Isthmus.[7]

In 1870 and 1871, Commodore Shufeldt, under the direction of the Secretary of the U. S. Navy, made a survey of the Isthmus "to ascertain the practicability of a ship-canal between

---

[1] Senate Ex. Doc. No. 25, Thirty-ninth Congress, Second Session, p. 6.

[2] Ditto, page 9. Also Report of Secretary of Finance of Mexico of the 15th of January, 1879, p. 52.

[3] Ditto, p. 52.   [4] Ditto, p. 52.

[5] Laws of Vermont for 1868, p. 218.

[6] Report of Secretary of Finance of Mexico, p. 52.   [7] Ditto.

the Atlantic and Pacific Oceans." [1]   He reported it practicable,
as a matter of engineering.[2]

In 1871, the La Sere grant became forfeited for non-fulfil-
ment of conditions.[3]

On the 22d of May, 1872, the grants of December 28, 1868,
and December 14, 1870, were renewed by Mexico.[4]

On the 15th of January, 1874, the same grants were again
renewed by Mexico.[5]

On the 14th of December, 1874, the said grant, for purposes
of a canal, was amended by the Mexican Congress, giving "new
and important advantages to the Company." [6]

Early in October, 1878, application was made to the Mexican
Government in behalf of Edward Learned of Massachusetts, for
a railway grant across the Isthmus.   On the 31st of the same
month it was favorably acted upon by the Executive of Mexico,
and referred to the proper committee of Congress.   They
made a favorable Report, which was approved by the Chamber
of Deputies by the significant vote of one hundred against
only twenty-eight, showing, as Secretary Romero well said,
"the feeling of a majority of the Chamber of Deputies with
regard to railroad grants in favor of North American Companies." [7]

On the 31st of May, 1879, President Diaz, acting under the
authority of an Act of Congress, declared all prior grants at
the Isthmus to be null and void.

On the 2d of June, 1879, the Mexican Congress completed
and ratified the pending Executive grant to said Learned, con-
ceding to him, or such company as he might organize, the
right to construct a telegraph line and railway across the
Isthmus.[8]

---

[1] Senate, Ex. Doc. No. 6, Forty-second Congress, Second Session.
[2] Ditto, p. 20.   [3] Ditto.
[4] Report of Secretary of Finance of Mexico, p. 53.
[5] Ditto, p. 53.   [6] Ditto, p. 53.   [7] Ditto, p. 53.
[8] See pamphlet of the Company, entitled "The Tehuantepec Railway
Grant."

In pursuance of this grant "The Tehuantepec Inter-Ocean Railroad Company" was organized, November 18, 1879, under the general railroad law of the State of Massachusetts, and Edward Learned elected President.

The Company have their head-quarters in New York, have recently purchased and shipped to the Isthmus rails for one-third of the line, are now engaged in the work of construction, and will soon consummate the grand enterprise which the indomitable Hernando Cortez commenced three hundred and fifty years ago.

CHAPTER II.

Descriptive Notes.

THE Isthmus of Tehuantepec is near the southern extremity of Mexico, and comprises the eastern portions of the states of Vera Cruz and Oaxaca. Its width in the narrowest part and in a direct line from ocean to ocean is 143¼ miles.[1] Its length, east and west, is 106 miles, and is terminated by two parallel lines, as indicated by the map on the preceding page. We have been unable to find the length definitely stated in any survey of the Isthmus, or in any geographical dictionary, notwithstanding its divisions, towns, products, and other features are fully described. It is probably because the Isthmus is so shaped that its terminal lines may, with propriety, be differently located. In one popular publication we find its length on the Gulf side defined as extending from the Bay of Alvarado to the State of Yucatan. That of course is too loose a construction of the term.

But in the Report of the principal survey is a map of the Isthmus which apparently includes nothing else, and we have followed its lines in the map we have prepared. It evidently coincides with the area covered by the several towns mentioned, and with the region described in the statement of products.

The area of the Isthmus, according to this definition, is over 15,000 square miles, or larger than the combined areas of Massachusetts, Rhode Island, and Connecticut.

[1] Barnard and Williams' Survey, p. 57.

It has three distinct divisions, which are clearly and concisely described by J. J. Williams, whose Report of the Survey of 1850-1 is the leading authority on this Isthmus, and is largely followed by most recent writers. He says as follows : "In considering the Isthmus with reference to its general topographical features, it may properly be said to comprise three main divisions, more or less distinct in their general characteristics, the first embracing that portion extending from the Gulf to the base of the Cordillera, and which may be called the *Atlantic plains;* the second comprising the more elevated or *mountainous districts,* in the central parts, and the third including the level country bordering the ocean on the south, and known as the *Pacific plains.*" * * * "The first division comprises a belt of country of some forty or fifty miles in breadth, lying contiguous to the Gulf coast, and made up of extensive alluvial basins of exceeding richness and fertility, through which the drainage of the northern slope of the Cordillera discharges itself into the Gulf." * * * "The second or middle division may be said to extend from the Jaltepec River on the north to within twenty or twenty-five miles of the Pacific coast, comprising a strip of country through the central portions of the Isthmus of some forty miles in breadth on the west, and gradually widening out towards the east to sixty or seventy miles. This division represents a great diversity of feature." * * * "By a narrow opening in these mountains we descend suddenly from the elevated table-lands to the Pacific plains which form the third or southern division. These plains average about twenty miles in breadth from the base of the mountains to the Pacific coast, and descend on the meridians to the lagoons at an inclination varying from ten to fifteen feet in the mile, thus forming, as it were, an immense inclined plane, with its side next the mountains, about 250 feet above the Pacific."[1]

By referring to the map of the Isthmus the reader will observe that the middle division, or *mountainous district,* is in-

[1] Barnard and Williams' Survey, pp. 13–17.

dented on the north by a lower elevation in the shape of table-lands, which extend southwardly from the rivers Jaltepec and Chalchijapa nearly across its whole width. These table-lands are about 1,400 square miles in extent, or larger than the State of Rhode Island.[1]

Between them and the *Pacific plains* are six known passes through the higher elevations, or southern part of the middle division.[2]

Their names and elevations above the level of the Pacific Ocean are given as follows in Williams' Report:[3]

La Chivela	780 feet
Masahua............................	843 "
West Piedra Parada.................	800 "
East Piedra Parada.................	825 "
Tarifa.............................	684 "
Convento	750 "

By again referring to the foregoing map, or what is better, a drainage map of the Isthmus, the reader will find, what is very unusual in Mexico, an abundance of rivers, a few of which are navigable. To their presence is largely due the wonderful fertility which we will have occasion to describe on subsequent pages.

Climate and Health.

In his valuable Report on the Survey of 1851, Williams, after reviewing in detail the subject of climate, reaches the following conclusion : " The conviction in the minds of those engaged in drawing up this Report, and one founded on a residence upon the spot, is that the climate of the Isthmus is a mild and healthy one, favorable to longevity, and free from many diseases incidental to more temperate latitudes. The health of those engaged on the survey was unusually good during their entire stay ; and although frequently, by accidents, wetted to the skin,

---

[1] Barnard and Williams' Survey, p. 15.   [2] Ditto, p. 81.   [3] Ditto, p. 81.

and remaining in wet clothes the whole day, and this occurring on successive days, with limited food at long intervals, yet none suffered in consequence—a strong proof that their health was due to the favorable climate."[1]

Doctor Kovaleski, one of the surgeons who accompanied this surveying party, testifies to the same effect in his report. Of the northern division of the Isthmus he says : "I took particular care to inquire among the inhabitants what were the diseases from which they most suffered, and how strangers settling among them were affected, and I ascertained beyond doubt that not only Minatitlan, but the whole plain of the Coatzacoalcos River, wherever inhabited, was a remarkably healthy country. Not a single case of yellow fever has ever occurred in Minatitlan, or any other part of the Coatzacoalcos plain, although in the years from 1829 to 1832, when the French emigrants attempted to form a colony on the Isthmus, the number of unacclimated strangers was considerable in the country, and they were exposed to every kind of privation and suffering."[2]

Of the middle division, or mountainous district, he says : "This entire region, for its salubrity, cannot be surpassed by any country whatever," and adds that three villages in the table-lands are patronized as health-resorts by people from Oaxaca and other Mexican states.[3]

Of the *Pacific plains* he continues : "Although the mean temperature of this valley is higher than that of the plain of Coatzacoalcos, it enjoys a degree of salubrity not inferior to that of the region of the mountains."[4]

In conclusion, he says of the whole Isthmus: "All these three regions together form a broad surface of country, from the Gulf of Mexico to the coast of the Pacific, of a great variety of resources and of remarkable healthiness, a feature peculiar to the Isthmus, as the lands on both of its sides are very unhealthy—such as Vera Cruz and Tabasco on the Gulf, Acapulco, Huatulco, and the coast of Guatemala on the Pacific shore.

[1] Barnard and Williams' Survey, p.172. [2] Ditto, p. 174.
[3] Ditto, p. 176. [4] Ditto, p. 177.

This peculiar and exclusive salubrity of the Isthmus is, in my opinion, chiefly due to its configuration, which forms, as it were, a gate, walled on both sides by heavy masses of mountains, through which pass currents of air, that render the country they traverse so permanently salubrious. That the winds prevail only within the limits of the Isthmus, and not within a few miles on either side of it, I am informed from most reliable sources."[1]

Surgeon J. C. Spear of the U. S. Navy, who was a member of Commodore Shufeldt's surveying party in 1870-1, testifies in his report: "It is the boast of the inhabitants that yellow fever has never visited them."[2]

Of the temperature he says: "Frost is never seen on the Isthmus, and none of the mountain peaks in sight from any point have snow on them, even in winter. The temperature in the summer never rises very high, rarely above ninety degrees."[3]

In a letter from Coatzacoalcos, dated August 17, 1880, Martin Van Brocklin, late chief engineer of the Metropolitan Elevated Railway Company of New York, and at present chief engineer of the Tehuantepec Railroad Company says: "The thermometer hanging in my office has not been above 85° since July 10, when it reached 88°, nor has it been more than 10° below those figures.

Products.

The popular idea of an inter-oceanic railway is usually associated with that of through freights. But those who have had occasion to observe how richly nature has endowed the Isthmus of Tehuantepec and consider its marvelous yield of nearly all the great staple products of the earth, see another source of profit largely supplementing that arising from transportation between the two seas. We think it easy to demonstrate that its local business per mile will, when properly developed,

[1] Barnard and Williams' Survey, p. 179.
[2] Shufeldt's Survey, p. 110.   [3] Ditto, p. 108.

equal that of the most profitable section of the New York Central Railroad between Albany and Buffalo. But facts best speak for themselves, and we will proceed to state them.

COFFEE.—Of this product, Mr. Williams says, in his official report: "The banks of the Coatzacoalcos exhibit in a wild state the greatest abundance of coffee, and, with few exceptions, no pains are taken to cultivate it, although the quality is admitted to be very superior. This neglect may be readily accounted for in the universal preference which exists among the natives for chocolate."[1]

U. S. Consul Hoyt, in a report to the State Department dated Minatitlan, Nov. 31, 1868, says, "Coffee attains perfection in three years, almost anywhere on the Isthmus."[2]

Hon. John W. Foster, late U. S. Minister at Mexico, in an elaborate review of Mexican coffee, published by the U. S. Department of Agriculture in 1876, said its quality was "equal to the best known in any country," and that Mexico in coffee alone "possesses a far greater source of wealth and prosperity" than in her product of silver.[3] To appreciate this tribute we should bear in mind that her yield of silver averages about $26,000,000 annually.[4]

Whoever develops this important industry upon the Isthmus will not have to go far to find a market; for the United States, during the ten years ending June 30, 1879, imported coffee to the amount of $452,615,511 in value, or $45,000,000 annually.[5] During the last one of those years, her imports were $47,356,819, of which $31,795,101 came from Brazil.[6]

As the Mississippi Valley is a great consumer of coffee, and as the mouth of its great river system is 4,550 nautical miles nearer the Isthmus of Tehuantepec than to Rio Janeiro, it should give the neighboring Isthmus the preference.

[1] Barnard and Williams' Survey, p. 188.
[2] Commercial Relations for 1868, p. 652.
[3] Monthly Report of Department of Agriculture for 1876, pp. 269–74.
[4] "The Silver Country," by A. D. Anderson, p. 43.
[5] Statistical Abstract No. 2 of Bureau of Statistics, p. 133.
[6] See Commerce and Navigation for 1879.

San Francisco purchases her coffee elsewhere than in Brazil. During 1879 she imported, in all, 13,889,462 pounds, of which nearly all, or 11,757,501 pounds, came from Central America.[1] In other words, to get her supply she passed directly by the door of Tehuantepec. When the Isthmus is opened by a railway, she will naturally buy at the nearer source of supply.

SUGAR.—The report of Commodore Shufeldt's Survey of 1870-1 says, " Sugar-cane grows in all the several natural divisions of the Isthmus, but the high ridges on the Atlantic Slope are best suited to its cultivation. It is said that a cane-field here produces as long as thirty years without replanting, when properly managed; and we saw, ourselves, several deserted plantations on the banks of the Coatzacoalcos where the cane was large and very sweet." [2] Still more emphatic testimony on this subject may be found in a letter to the State Department by U. S. Consul Hoyt, from Minatitlan, in 1868. After speaking of the then proposed railway across the Isthmus, and the American capital likely to flow in as a result of that project, he says : " The culture of sugar-cane I look upon as being the best investment for large capital, requiring but one year for maturing the first crop ; after which two crops can be obtained in fifteen months. The cane once started, will continue to yield prolifically twenty years without replanting ; and I have seen plantations that have not been replanted for forty years. When the plants have obtained a good setting, they require but little care or cultivation, as the towering growth of the stalk effectually shades the ground, and prevents the growth of weeds. I have seen a sample of sugar-cane eighteen feet in length and three inches in diameter. The yield will average a ton of sugar to the acre at each cutting after the first year, and the product of rum and molasses will be sufficient to pay the expenses of cultivation. The lands along the banks of all the rivers are admirably adapted to the culture of sugar-cane, and in many places on the high lands, I have seen good crops suc-

[1] San Francisco Journal of Commerce, Jan. 21, 1880.
[2] Shufeldt's Survey, p. 117.

cessfully raised year after year without irrigation, the rainy
season being sufficiently long to mature the crop." [1]

For this product also there is an excellent market in the
United States, her imports of sugar alone for the ten years ending June 30, 1879, being $703,200,567 in value, or $70,000,000
annually.[2]

ORANGES.—When one of the promoters of the proposed Florida Ship Canal was asked by a New York *Herald* correspondent
how the company proposed to improve the land grant of one
million acres it was seeking from the State of Florida, he re-
plied : "By colonizing Frenchmen and Germans upon it to
grow oranges and olives and the like products of Southern
France and Germany." [3] Tehuantepec possesses equal if not
better conditions for successfully producing this valuable fruit,
as may be seen from the following extract from the Report of
Shufeldt's Survey : "Oranges grow in all parts of the Isthmus,
but those of the Atlantic plains and the central division are the
best. At Santa Maria Chimalapa this fruit is particularly fine,
and there are two crops yearly, but not always on the same
tree." * * * "On the banks of the Coatzacoalcos there is
an orange grove of about ten acres, planted by an American,
where the fruit is equal to the best Havana oranges. We mea-
sured a large basketful, taken indiscriminately from the trees,
and found the average circumference of the oranges to be $9\frac{11}{16}$
inches. We estimated that the ripe crop we found on this deserted and untilled orange grove was as much as one hundred
and fifty bushels to the acre. On the Isthmus of Tehuantepec
there is no frost to blight this crop as there is occasionally in
Florida and Louisiana, nor hurricanes to destroy it, as in the
West Indies ; nor are the northers violent enough on the Atlantic plains to materially injure it. Oranges of the very best
quality can be readily grown along the banks of the Coatzacoal-

[1] Commercial Relations for 1868, pp. 651-2.
[2] Statistical Abstract No. 2 of Bureau of Statistics, p. 131.
[3] N. Y. *Herald*, August 2, 1879.

cos and Uspanapa in quantity sufficient to supply the market of the United States." [1]

Of the demand for oranges and lemons in the United States and Europe, the Annual Report of the U. S. Department of Agriculture for 1877 says as follows : " The statistics of the orange trade, both in this country and Europe, bear out the statement that but few valuable fruits ever exceed the demand." * * * " The importations into Great Britain show that the trade and consumption in these fruits have more than doubled in that country during the past ten years." * * * " The importation of oranges and lemons into France from Spain and Italy have increased in the last few years more than fourfold in quantity." [2]

BANANAS.—Doctor Spear, of Shufeldt's Survey, describes the great profusion of this product on the Isthmus as follows : " There are as many as fifteen well known varieties of the banana, some of which are of a very superior quality. The manzána variety is the most delicious of all, and, as its name indicates, has the flavor of the apple. Like the orange, the best bananas are found on the Atlantic plains and in the central division, and they are ripe at all seasons of the year." [3]

Williams' Report says of the same fruit : " The warm, humid valleys of the Gulf shore appear to be the natural position for the banana, where the fruit is occasionally eight inches in circumference, with a length of ten or twelve. Forty plants growing on a space of 1,070 feet are calculated to furnish 4,400 pounds of nutritive matter—a quantity above the acreable product of any cereal crop. In other words, the same extent of ground under bananas which will support fifty individuals, when under wheat will only support two." [4]

CORN.—Of this staple product Williams' Report says : " This is the native country of maize, and upon the wet land, *milpos* (those subject to periodic overflow), the yield is two crops an-

---

[1] Shufeldt's Survey, pp. 119–120.
[2] Report of Department of Agriculture for 1877, pp. 566–7.
[3] Shufeldt's Survey, p. 120.   [4] Barnard and Williams' Survey, p. 194.

nually, each of which averages sixty bushels to the acre, and without other labor than the mere planting. Indeed, it is no uncommon sight to see the reaper and sower in the same field." * * * "On the margin of the Uspanapa at some of the elevated points, *three* crops, each yielding *seventy* bushels to the acre, have been raised in a favorable year." * * * "The fecundity of the Mexican variety of maize is astonishing. Fertile lands usually afford a return of three or four hundred fold; even when the soil is sterile, the produce varies from sixty to eighty. The general estimate for the Isthmus may be considered as one hundred and fifty fold."[1]

In the United States, the average yield of corn per acre, in 1878, was $26\frac{2}{10}$ bushels, and, of course, but one crop per annum."[2]

COTTON.—Another important industry which can be successfully developed upon the Isthmus is cotton-growing. Williams' Report says: "The cotton-plantations of the Isthmus are so trifling as scarcely to deserve the name, but the fitness of the soil and climate to produce it are beyond question. There are two varieties, one of which, raised in the neighborhood of Minatitlan, is not inferior in texture, whiteness, or length of staple, to the finest uplands of the Southern United States." * * * "What would seem to favor the cultivation of cotton is the sheltered condition of the table-lands and savannas, and the entire absence of the *army-worm* which so seriously damages the cotton crops of the Southern States. It is entirely unknown to the natives."[3]

Mexico has for centuries been a large consumer of cotton. Mayer, the historian, says: "Cotton was among the indigenous products of Mexico at the time of the conquest ; and the early adventurers not only found it to constitute the common vesture of the masses of the people, but also that the most delicate and luxurious articles of dress were made of it. The Aztecs pos-

[1] Barnard and Williams' Survey, pp. 185-6.
[2] Statistical Abstract No. 2 of Bureau of Statistics, p. 157.
[3] Barnard and Williams' Survey, pp. 188-9.

sessed the art of spinning it to an extreme degree of fineness, and of imparting to it the beautiful and brilliant dyes for which they were celebrated." [1]

There were, in 1876, over seventy cotton-factories in that republic, most of them being in the states of Mexico, Puebla, • and Vera Cruz, not far distant from the Isthmus. They require annually 7,013,122 kilograms of raw cotton.[2]

Even these factories do not supply the Mexican demand, for the chief item of her annual imports is invariably cotton. Of her total imports for the fiscal year ending June 30, 1873, amounting to $29,062,406 in value, $10,531,970, or over one-third, were cotton stuffs.[3]

In this connection, it should be stated that the Isthmus possesses superior water-power, a description of which will be given on a subsequent page.

As one of its commercial outlets rests upon the Pacific, the Isthmus has excellent facilities for supplying the extensive cotton demand of China, Japan, and the western coast of Central and South America.

These facts and figures point to the Isthmus as a most promising field for cotton-growing, and capitalists would do well to give the subject careful consideration.

Tobacco.—Doctor Spear, of Shufeldt's Survey, reports: " The best tobacco found in Mexico grows on the Isthmus of Tehuan-tepec, on the Atlantic plains west of the Coatzacoalcos River. It is extensively cultivated by the Indians belonging to the towns of Jaltipan, Acayucam, and Chinameca." [4]

The Report of the prior survey, by Williams, says: " The plantations of tobacco (the *Nicotiana tabacum, Linnæus*) are both numerous and considerable, especially in the northern and central divisions of the Isthmus. That raised in the Chimalapas and on the uplands generally, is known by the name of " Tabaco del Monte." This variety is powerfully narcotic,

[1] Mexico ; Aztec, Spanish, and Republican, by B. Mayer, vol. ii, p. 67
[2] The Republic of Mexico in 1876, by A. G. Cubas, p. 28-9.
[3] Commercial Relations for 1875, p. 1129.
[4] Shufeldt's Survey, p. 118.

coarse, and grows to a large size, the leaves averaging *thirty-three* inches in length, and *fifteen* in breadth.

Another kind cultivated on the plains, and called "corral," is smaller, and of a flavor and quality which is said to be superior to the best *vuelta de abaja* of Cuba."[1]

Cocoa.—This product is raised chiefly in the northern division of the Isthmus, and is not found on the Pacific plains. Williams reports two qualities, and of one he says : "The quality is said to be superior to that of Guayaquil or Maracaibo, and the prolific return which characterizes its cultivation is sure evidence of its importance and value. The lands east of the Coatzacoalcos seem particularly adapted to its growth."[2]

Rice.—U. S. Consul Hoyt reports that "rice grows well either in the high or low lands, the best quality and yield being obtained on the mountains dividing the Atlantic and Pacific coasts."[3]

The Report of Shufeldt's Survey states : "Rice at present is produced in only very small quantities, but as soon as there is a demand for it, the central division and the Atlantic plains will be found capable of producing it in any quantity that may be desired."[4]

On this subject Williams quotes from Don Tadeo de Ortiz, who was appointed by the state of Vera Cruz to survey the Isthmus in 1824. In speaking of the region between the Coatzacoalcos and Tonala rivers, he says : "That which most particularly characterizes this privileged region is the singular fact that one single sowing of rice will yield successively two large crops without the slightest additional labor."[5]

Timber.—After reporting upon various natural products of the Isthmus, Williams says : "But among all these productions the timber of its immense forests deserves particular mention. Its abundance is such that the only limit which can be assigned to the supply it may yield is the demand for centuries to come. From the fir, the oak, the cedar, and every description of build-

[1] Barnard and Williams' Survey, pp. 187-8. [2] Ditto, p. 187.
[3] Commercial Relations for 1868, p. 650. [4] Shufeldt's Survey, p. 117.
[5] Barnard and Williams' Survey, p. 188.

ing timber, to the dye and fine woods, their profusion is absolutely incredible. None of the countries which at present supply these species of woods could compete with the Isthmus, where they are found on the very banks of the river which facilitate their carriage." [1]

The great variety of valuable timber and fine woods may be seen from the enumeration in Shufeldt's Survey, as follows : " In the Chimalapa region, along the banks of the Upper Coatzacoalcos and the Rio Blanco, white pine, pitch pine, white oak, cypress, mammee-sapote, and chico-sapote abound." * * * "On the Atlantic plains mahogany, cedar, macaya, guapaque, mammee-sapote, and piqui, or iron wood, are abundant." * * * "On the Pacific plains lignum-vitæ, rosewood, calabash, and ebony are the most common." [2]

Robert Dale, who visited the Isthmus about 1850, and has since published, in London, a book on the excursion, says : "We found the mahogany trees very plentiful, growing generally from fifty to two hundred feet apart ; many of them were of great girth, say from eighteen to twenty-four feet in circumference, and as high as fifty or even sixty feet without branches, when they threw out large arms, often thickly covered with parasitical plants and flowers, and with vines hanging down from them. Here we saw a raft of one hundred logs on the river, besides other trees which were cut in the forests ready for removal. A good many of them, very sound and free from defects, were well squared, and about three feet deep." [3]

A work on the mahogany tree by Messrs. Chaloner and Fleming of Liverpool, is authority for the following : "The first mention of it occurs shortly after the discovery of the New World, when Cortez and his companions, between the years 1521 and 1540, employed it in the construction of the ships which they built for prosecuting their voyages of discovery after

---

[1] Barnard and Williams' Survey, p. 140.

[2] Shufeldt's Survey, pp. 121-2.

[3] Notes of an Excursion to the Isthmus of Tehuantepec, by Robert Dale, p. 34

their conquest of Mexico."[1] If these writers are correct in
this historical point, it is likely the mahogany for these ships
was cut at the Isthmus; for a part of Cortez's expedition to the
South Sea was fitted out at the port of Tehuantepec, or his own
port, as it is called.

The imports of mahogany by Great Britain during the ten
years ending 1879, were in value $22,200,102, or $2,220,000 annually.[2]

Of the varieties of fancy woods which, as we have above
stated, grow upon the Isthmus, the United States imported
from various sources during the year ending June 30, 1879, as
follows, in value:[3]

Cedar	$192,688
Mahogany	182,283
Rose	106,811
Ebony	43,055
Lignum-vitæ	19,781

Whoever makes the business connection between the enormous supply of these valuable timbers to be found in the great
forests of the Isthmus, and the steady demand existing in
Great Britain, the United States, and other countries, may feel
assured of large returns.

INDIA-RUBBER.—Williams' Report states that the India-rubber
trees are "found in astonishing numbers . throughout the
forests that skirt the tributary streams," and adds: "Taking
half the number of trees found within an area of one-fourth of
a square mile on the Uspanapa River as a basis of an estimate,
and allowing none to grow on the Pacific plains, there would
be found not less than two millions of India-rubber trees within
the limits of the Isthmus, some of which yield four and five
pounds of gum in a year. If, from this prodigious number of

---

[1] "The Mahogany Tree," by Chaloner and Fleming of Liverpool, p. 87.
[2] Statistical Abstract of United Kingdom, No. 27, pp. 48–9.
[3] Commerce and Navigation for 1879, p. 599.

trees, we suppose *one-half* only to be available, and that a single pound per tree per annum be the average yield, we should then have one million of pounds, which, at the present value of forty cents, would realize the sum of $400,000 for this article alone." [1]

The United States alone imported during the fiscal year ending June 30, 1879, India-rubber and gutta-percha, crude, to the value of $3,296,766.[2] Add to this the value of the imports of European countries, and it is apparent that this product of the Isthmus, when properly developed, will find a ready market.

VEGETABLE DYES.—Of these products, Williams reports: "The enumeration of all the vegetable dyes found on the Isthmus, with all that might be said of the numerous varieties, would constitute matter for a well-filled volume on botany, rather than the general details of a statistical report. Many, nevertheless, deserve an especial notice, either from the brilliant colors which they yield, or the pecuniary considerations involved in their growth and production. Among these may be classed the indigo-trees, which are indigenous to Mexico." [3]

INDIGO.—Of this dye, Shufeldt's Report says : "The most important article of export cultivated on the Isthmus is indigo. Its cultivation is confined entirely to the Pacific side, in the light and dry atmosphere of which this plant seems to do remarkably well." [4]

The official representative of the United States, in a report to the State Department, dated Tehuantepec, October 15, 1871, says : "Indigo is the principal produce of this part of the country on the Pacific plains, from twenty to forty miles wide in a direct line from the sea to the Cordilleras, and about eighty miles in length, running nearly east and west." * * * "If an inter-oceanic canal should be constructed across the

[1] Barnard and Williams' Survey, pp. 183–4.
[2] Commerce and Navigation for 1879.
[3] Barnard and Williams' Survey, p. 189.
[4] Shufeldt's Survey, p. 118.

Isthmus of Tehuantepec, the natural influx of foreign population drawn by such an enterprise would develop in a few years
the growth of indigo to an extent sufficient to rival the East
Indies, and increase many other products in corresponding
proportions."[1]

The imports of indigo by the United States during the
year ending June 30, 1879, were $1,488,481 in value, of which
about half, or $700,318, came from the distant British East
Indies.[2]

COCHINEAL.—Williams says: "Formerly this insect was reared
with care, and formed a valuable article of commerce until the
discovery of French chemical dyes."[3]

How extensively it formerly figured in the commerce of
Oaxaca, the state which includes the larger part of the Isthmus,
is told by the Acting British Minister who resided in Mexico
about 1827. He says of that state : "By the official returns
which I possess, it appears that the value of cochineal entered
upon the books of this office up to 1815, was $91,308,907,
which upon fifty-seven years, gives an average of $1,601,910
per annum, without making any allowance for contraband."[4]
The historian Mayer brings the record down to a later date
as follows : "It appears that from 1758 to 1832 inclusive, or
in seventy-five years, 44,195,750 pounds of cochineal were
produced in the state of Oaxaca alone, which were worth
$106,170,671 at the market price."[5]

We see no good reason why Oaxaca, when supplied with a
railway across the Isthmus, should not regain some of her lost
prestige in this industry; for the United States, during the
year ending June 30, 1879, imported cochineal to the value of
$716,171. Of this total, only $41,931 was from Mexico, the
rest coming from England, Africa, and adjacent islands.[6]

[1] Commercial Relations for 1871, p. 916.
[2] Commerce and Navigation for 1879.
[3] Barnard and Williams' Survey, p. 215.
[4] Mexico in 1827, by H. G. Ward, vol. ᵏ pp. 84–5.
[5] Mayer's History of Mexico, vol. ii.
[6] Commerce and Navigation for 1879.

PETROLEUM.—In his official report from Minatitlan, in 1868, U. S. Consul Hoyt says : "Petroleum is sufficiently abundant in this district to supply the world. Indications of its locality exist everywhere, and in many places it comes to the surface and forms small lakes and springs to such an extent, that it can be dipped up in large quantities. In fact, the whole of this side of the Isthmus is a vast lake of petroleum, in my estimation ; and from the explorations I have made, I believe it can be found almost anywhere. Its richness has been tested by some of the best chemists in the United States. Professor Percy of New York has made an analysis of some of this petroleum taken from the surface (or rather from one of the springs), the result being as follows :

Volatile combustible matter........................... 75.95
Coke... 24.05
                                                 ―――――
                                                 100.00

Submitted to distillation in solid retorts it furnishes a thick lubricating oil of specific gravity of 0.897 (26° Beaumé) in the following proportions : One pound of bitumen yields eight and a half fluid-ounces of oil ; hence one ton of 2,240 pounds yields 148.75 gallons, or 3.72 barrels of forty gallons each.

A separate fractional distillation, with the heat raised to 400° Fahrenheit, yielded, in hundred parts :

Illuminating oil..................................... 13
Lubricating oil...................................... 87
                                                 ―――
                                                 100" [1]

IXTLE OR PITA.—Williams' Report of the Railway Survey says : "Among the spontaneous products is the *bromelia pita,* or ixtle of the Isthmus, which differs in some respects from the *agave americana* of Europe, the *pulque de maguey* of Mexico, and the *agave sisalana* of Campeachy. Of this prolific plant there are numerous varieties, all yielding fibres which vary in quality from the coarsest hemp to the finest flax. Nor is the value of

―――――――――――――

[1] Commercial Relations for 1868. p. 652.

the plant diminished by its indifference to soil, climate, and season. The simplicity of its cultivation, and the facility of extracting and preparing its products, render it of universal use. From it is fabricated thread and cordage, mats, bagging, and clothing, and the hammocks in which the natives are born, repose, and die. The fibres of the pita are sometimes employed in the manufacture of paper; its juice is used as a caustic for wounds, and its thorns serve the Indians for needles and pins." [1] On the following page he says: "In 1831, according to the statements of Señor Iglesias, the ixtle plantations in the northern division numbered 1,221." [2]

In a recent conversation with a correspondent of the New York *World*, Dr. Trowbridge, the U. S. consul at Vera Cruz, in speaking of this product as found in Southern Mexico (either Vera Cruz or Yucatan), said: "There is a species of cactus here called pita (I do not know its botanical name), some of the fibres of which are sixteen feet long. It is strong and silky, and capable of being divided into threads, from which gossamer webs might be woven. In fact a few months ago a Vera Cruzano sent some of the fibre to England and had a few handkerchiefs made, which were extremely beautiful and appeared more like silver tissue than linen, and were quite strong." * * * "There are millions of acres of the pita plant, which, if utilized, might bring an immense revenue to the coffers of this impoverished country." [3]

CATTLE.—We have stated on a previous page that Cortez had large estates on the Isthmus. Part were in the middle division and part on the Pacific plains, near the town of Tehuantepec. On his return from Spain to Mexico, in 1530, says the historian Prescott, "he imported large numbers of merino sheep and other cattle, which found abundant pastures in the country around Tehuantepec." [4] Three hundred and twenty years later, or in 1850-1, the same grazing facilities were found by

[1] Barnard and Williams' Survey, p. 184.   [2] Ditto, p. 185.
[3] N. Y. *World*, May 3, 1880.
[4] Prescott's Conquest of Mexico, vol. iii, p. 332.

Williams. He reports : "The immense potreros which border all the principal streams on the northern division, furnish rich pastures of never-failing verdure for numerous herds. During the short season that these potreros are inundated, the cattle are driven to the more elevated savannas remote from the river margins. The extensive table-lands in the central portions of the Isthmus, as well as the plains bordering the Pacific, also furnish abundance of excellent pasturage. Indeed the whole country seems peculiarly well adapted to the raising of horned cattle. With little care on the part of their owners, they in-crease rapidly, grow to a large size, and have a remarkably sleek and well-favored appearance. Enjoying a range of the finest pastures in the world, they are usually in good condition and make fair beef." [1]

FISH AND GAME.—According to the same Report, "deer are found in great multitudes in all sections, and serve as an abun-dant source of prey for the numerous voracious animals which infest the country." [2] Also, "there is perhaps no country in the world, situated within the same parallels of latitude, that pro-duces an equivalent quantity and variety of fish and wild game as the Isthmus." * * * "Excellent fish are found in great pro-fusion in all the rivers and arroyos that drain the slopes of the Cordillera, particularly in the smaller streams. In most of the larger ones there are many varieties of good size, and fine quality ; indeed, fish constitutes an important item of food for the inhabitants. Those living at Santa Maria Chimalapa, having but few domestic animals, and no means of killing game, subsist almost entirely upon the fish obtained from the Rio del Corte. These are taken in such numbers that they are salted and transported to supply the towns of the central division." [3]

The great number of these fish-producing streams may be seen by a glance at a map of the Isthmus.

[1] Barnard and Williams' Survey, p. 202.   [2] Ditto, p. 207.   [3] Ditto, p. 212.

Water-Power and Manufactures.

By referring again to a drainage-map, the reader will readily see that nature has provided abundant water-power for manufacturing the raw material of which the Isthmus is so prolific. These many streams, both on the Atlantic and Pacific slopes, have their sources in the central, or mountainous district, and therefore have a sufficient fall to furnish a motive-power.

Of the rivers on the Pacific slope, Williams reports : " All of these streams, as they issue from the mountains, are remarkably pure and limpid, even in times of flood, thereby indicating the rocky nature of the districts which they drain. In their descent towards the plains, they offer almost unlimited sources of water-power."[1] His description of the leading rivers of this slope also indicates very clearly that the supply of water is unfailing, for he says : "The elevated mountain-peaks, near the source of these streams, are almost constantly enveloped in clouds, which may account for their remarkably uniform flow of water throughout the entire year."[2]

Doubtless the same may be said of many of the rivers of the Atlantic slope, for their sources are in the same dividing ridge.

Another important consideration is that these streams are never frozen, as is occasionally the water supply in manufacturing districts of New England.

All of the conditions are then favorable for developing at the Isthmus an extensive manufacturing industry. Factories adjoining the cotton-fields and pita-plantations, and saw-mills in the great forests of mahogany, cedar, and other valuable woods, may here be made a source of great profit.

[1] Barnard and Williams' Survey, p. 18.   [2] Ditto, p. 18.

Population.

Those unfamiliar with the ancient and modern history of the Isthmus must not think that it is entirely undeveloped. At the time of the railway survey by Barnard and Williams in 1851, there were on the Isthmus thirty-eight towns and cities with a total population of 61,393 souls.[1] The city of Tehuantepec, on the Pacific side, was the largest, and contained 13,000 people. In the list is another containing over 8,000 inhabitants, another containing 6,000, another 5,000, three containing 2,000 each, and seven having 1,000 each.

The Report of the Canal Survey of 1871 says : "The last census of the Isthmus of Tehuantepec gives the number of inhabitants as 82,395 ; but since this enumeration was made, the population, from civil wars and other causes, has greatly fallen off, and probably does not exceed 50,000."[2]

Formerly Southern Mexico was the most densely populated part of that republic, and with railway facilities to develop its visible and latent resources, its future population is likely to exceed that of the native races found there by Cortez, and possibly equal that of pre-historic times, when, as M. Charnay states of a part of Yucatan, east of the Isthmus, "there are unmistakable indications that this district, and these cities, were more densely populated than any known portion of the globe at the present moment."[3]

Antiquities.

A little east of the Isthmus, in the state of Chiapas, are the massive ruins of the ancient city of Palenque, which Stephens, the explorer, says "are the first which awakened attention to the existence of ancient and unknown cities in America."[4]

[1] Barnard and Williams' Survey, p. 257.
[2] Shufeldt's Survey, p. 135.
[3] Remarks on Lorillard Expedition in New York *Herald*, April 22, 1880.
[4] Incidents of Travel in Central America, Chiapas, and Yucatan, by J. L. Stephens, vol. ii, p. 294.

The principal building, called the Palace, is thus described by
Baldwin : "It stands near the river, on a terraced pyramidal
foundation forty feet high, and three hundred and ten feet
long, by two hundred and sixty broad, at the base. The edifice
itself is two hundred and twenty-eight feet long, one hundred
and eighty wide, and twenty-five feet high. It faces the east, and
has fourteen doorways on each side, with eleven at the ends.
It was built entirely of hewn stone, laid with admirable pre-
cision in mortar, which seems to have been of the best quality.
A corridor nine feet wide, and roofed by a pointed arch, went
round the building on the outside, and this was separated from
another within, of equal width. The "Palace" has four interior courts, the largest being seventy by eighty feet in extent.
These are surrounded by corridors, and the architectural work
facing them is richly decorated." [1]

A clearer idea of the grandeur of this palace, and of other
ruins of Palenque, may be found in Lord Kingsborough's costly
illustrations of the "Antiquities of Ancient Mexico," vol. iv,
Plates 19 to 45 inclusive, of Part III ; in Catherwood's "Views
of Ancient Monuments," Plates 6 and 7 ; in Stephens' illustrated work on "Chiapas, Yucatan," etc., vol. ii, pp. 284–358,
and in Bancroft's "Native Races," vol. iv, p. 310, which contains a view of the palace restored.

Stephens' comment, after visiting these ruins, indicates the
attractions they offer for the student of Ancient America. He
says : "In the romance of the world's history nothing ever
impressed me more forcibly than the spectacle of this once
great and lovely city overturned, desolate, and lost." [2]

A little west of the Isthmus, in Oaxaca, are the ruins of the
ancient city of Mitla. They are even more magnificent than
those at Palenque. According to Baldwin, the ruins of six
edifices and three pyramids are all that now remain.[3] He gives
a translation of Dupaix's description of four palaces, as follows :
They "were erected with lavish munificence ;" * * * "they

[1] Ancient America, by J. D. Baldwin, p. 104.
[2] Vol. II of his " Incidents of Travel," p. £57.
[3] Ancient America, by J. D. Baldwin, p. 117.

combine the solidity of the works of Egypt with the elegance of those of Greece." * * * "But what is most remarkable, interesting, and striking in these monuments, and which alone would be sufficient to give them the first rank among all known orders of architecture, is the execution of their mosaic relievos, very different from plain mosaic, and consequently requiring more ingenious combination and greater art and labor. They are inlaid on the surface of the wall, and their duration is owing to the method of fixing the prepared stones into the stone surface which made their union with it perfect." [1]

The justice of this tribute will readily be admitted by the reader, if he will visit the Congressional Library at Washington, the Astor Library in New York, or some other of the few public libraries which possess a copy of Lord Kingsborough's illustrated work, and examine Plate No. 82 of vol. iv, Part III, which represents the ruins of the "First Palace."

Other illustrations of the ruins of Mitla may be found in Bancroft's "Native Races," vol. iv, pp. 388–417 ; and in vol. iv, Part II, Plates 78-95, inclusive, of Lord Kingsborough's work.

East of the Isthmus, in Yucatan, the evidences of a dense population and high civilization are very numerous.

Bancroft says : "The state is literally dotted, at least in the northern central, or best known portions, with ruined edifices and cities." [2] Stephens says of the same state : "In our long, irregular, and devious route, we have discovered the crumbling remains of forty-four ancient cities, most of them but a short distance apart." [3]

Of the Isthmus itself, Williams reports : "Everywhere on the Isthmus—even on loftiest mountains, in the deepest dells, and in the most impenetrable forests—there are silent evidences of the history of a vast and powerful people, of which there scarcely remains now a tenth part." [4] In speaking of the region around the city of Tehuantepec, on the Pacific plains, he says :

[1] Ancient America, by J. D. Baldwin, p. 117.
[2] "Native Races," vol. iv, p. 143.
[3] His "Incidents of Travel," vol. ii, p. 444.
[4] Barnard and Williams' Survey, p. 223.

"One of the most interesting features near Tehuantepec is Mount Guiéngola, some five leagues distant in a northwesterly direction. This mountain is celebrated for having once been inhabited by a very large population, the evidences of which are palpable to this day from the immense heaps of ruins which are now found in various parts of it."[1]

In Lord Kingsborough's illustrations of Ancient Mexico, Plates 1 to 18 inclusive, of vol. iv, Part III, are devoted to the "Antiquities of Tehuantepec." One of these represents the pyramid near the city of Tehuantepec.

It is hoped that the Lorillard expedition, under the direction of the distinguished M. Charnay, which a few months ago commenced a thorough exploration of Southern Mexico for the discovery of its antiquities, will give to the world more definite information about those on the Isthmus. The programme of this expedition, which is now at work near the city of Mexico, was recently announced in the New York *Herald* as follows : " Working its way around by Oajaca, the expedition will investigate the sculptures of Mount Alban, and the rich remains of that region. It will then proceed to Mitla, and after spending some time in the mountainous and almost unexplored districts adjoining the Tehuantepec, will arrive at Palenque. It is anticipated that this famous seat of worship will yield a rich store of inscriptions and bas-reliefs,"[2] etc., etc.

The growing interest in the history and antiquities of the civilized races of Ancient America is likely to make the Tehuantepec Railway a base of operations for tourists and explorers, for it is projected midway between Palenque and Mitla, which ancient cities are situated in regions likely to have tributary roads to connect with the trunk line across the Isthmus.

[1] Barnard and Williams' Survey, pp. 253–4.
[2] N. Y. *Herald*, June 20, 1880.

Harbors and Railway-Line.

The road commences at the mouth of the Coatzacoalcos River, on the Gulf of Mexico.

As we have stated in the "Historical Notes," Cortez sent a party of experts to examine the harbor at that point, was greatly pleased at their favorable report, and so expressed himself in a dispatch to the King of Spain.

Engineer Fuertes, who accompanied Commodore Shufeldt's Survey of 1870-1, reports : "I feel no hesitation in asserting with unusual confidence that the Coatzacoalcos River can be made a very safe and snug harbor for any class of ships with but comparatively small expense." [1]

The consulting engineer of the Tehuantepec Inter-Ocean Railroad Co., Hon. Wm. J. McAlpine, has, under date of Oct. 22, 1880, favored the writer with the following notes on the subject of harbors and location of the road, viz.: " During the early months of this year I made a thorough personal examination of the northern half of the Isthmus, and also of the harbor and river of Coatzacoalcos, for seventy-five miles of its length.

Previous and subsequent to these personal examinations, I have carefully studied all of the reports, surveys, and maps which have been made for this railroad, so as to form very correct opinions of the facilities and difficulties of constructing it, and its necessary harbors.

The following remarks are the result of these examinations and studies.

There are three distinctive districts of country through which this railroad must be built, viz. :

The northern table-lands, the middle (so called) mountain-district, and the southern, or Pacific plains.

The northern division of the railroad, of seventy-five miles length, commences on the Gulf of Mexico, at the mouth of the Coatzacoalcos River, and, after running ten miles on nearly a level grade, ascends to and follows on the eastern divide of

---

[1] Shufeldt's Survey, p. 36.

that river and the San Juan, until it descends to the Jaltipec River, a tributary of the Coatzacoalcos, which, running easterly, skirts the northern base of the Cordilleras.

The grades up to and over this divide are easy ; in no place exceeding forty feet to the mile. The highest part of the divide is from two hundred and fifty to three hundred feet above the level of the sea, and the crossing of the Jaltipec at the seventieth mile is but one hundred and twenty-five feet above the sea.

This division of the road presents no unusual difficulties ; only two streams of any size are crossed, and these require but single span bridges of one hundred and fifty feet each.

The dense tropical forests require heavy clearing, but the grading will generally be quite light.

The tributaries of the Coatzacoalcos penetrate the mountain region to within ten miles of the southern or Pacific base of the Sierra, and allow the location of the railroad along their valleys, with grades in no place exceeding fifty feet to the mile, and with the highest summit between the oceans of only seven hundred and seventy feet above the sea.

The line passes over several mountain plains where the grading will be light, and elsewhere it will not be expensive. The crossing of a dozen or so mountain streams will be made with bridges of spans of one hundred to one hundred and fifty feet at each place.

The western slope of the Cordilleras is steep, falling about five hundred feet in a direct line of four or five miles, but by extending the length of the line along the mountain slope it is believed that grades of eighty feet or less, can be obtained without an excessive cost.

The whole length of the mountain division will be about sixty miles.

The Southern, or Pacific division, will be from thirty to forty miles long, depending upon the location of the southern terminus. It extends from the southern base of the mountain over a gently sloping plain to the ocean, with easy grades, none of which will exceed thirty to forty feet to the mile.

The grading will generally be light, with no large water-courses to cross."

*   *   *   *   *   *

" The harbor on the Gulf side is one of the best on the Atlantic side of the continent south of Norfolk.

As soon as vessels have passed in from sea, they are completely protected from all winds from the ocean for half way around the circle, by the coast range of hills, which are from two to four hundred feet high, while on the land side the gigantic forests protect the harbor for the remainder of the circle.

I witnessed several severe storms, both from the ocean and landward side, and never saw the waters of the harbor as much disturbed as they are along the wharves of New York.

The water inside of the bar is from thirty to sixty feet deep, and extends more than twenty miles at not less than twenty-five feet depth and a width of one or two thousand feet, with many deep side *bayeux.*

The water on the bar was sixteen feet deep when first visited by the Spaniards, and has never changed in direction or depth.

This is due to the material which forms the bottom and sides of the channel, which is a hard, stiff clay, the same which is found along the bed and banks of the harbor and river, on which the swift currents have almost no visible effect.

There is a little sand overlying this bed of clay near the ocean shore, but a comparatively inexpensive mole will prevent it from drifting into the entrance channel, and the large volume of water from the six to eight thousand square miles of drainage of the Coatzacoalcos River, aided by the influx and reflux of the ocean water, will doubtless prevent any deposit being made in the new channel across the bar.

A cut of three thousand feet in length, averaging less than six feet depth, will open a straight channel of twenty-five feet depth directly into the calm water of the land-locked harbor.

I have examined the charts of all of the proposed harbors on the Pacific coast suitable for the terminus of the railroad, and particularly a very recent and accurate chart of several of the proposed harbors, and can confidently state that a most suitable one exists, which, with a moderate expenditure, can be adapted to immediate wants, and can be enlarged and improved from time to time, to meet the increasing demands of the growing commerce."

Land Grant.

That a grant of land through the rich agricultural regions which we have above described must become a source of immense profit to the Company is very apparent.

What then is the grant?

Article 10 says as follows : " Of unoccupied public lands the Government gives the Company such a strip as it may require for the line of the road, and, in addition thereto, one-half of unoccupied public lands that may be found within one league from each side of the railroad only, throughout the whole extent thereof ; *provided* that only such lands shall be considered unoccupied as shall not have been conveyed to private parties prior to or at the date of the approval of this contract, or taken up according to law prior to or at the same date."

It is estimated by the Company that the area thus granted will amount to about two hundred thousand acres.

Less than 1000 ft. above Sea.

1000 to 3000 " "

Over 3000

Existing Railways (year 1879)

CHAPTER III.

"No one can fail to see how exceedingly important this communication would be to the Government of Mexico. It proposes to give her a practical highway from sea to sea. It opens to her a communication on the one side and the other with the eastern and the western world. It gives her access to the markets of all nations, and makes her, in short, a central point of the commerce of modern times."

From official letter by DANIEL WEBSTER, *while Secretary of State, in 1851. See Sen. Exec. Doc. No. 97, Thirty-second Congress, First Session.*

Need of Railways.

THE United States, with an area of 3,026,494 square miles, had, in 1879, 86,497 miles of railways.[1]

Mexico, with an area of 763,804 square miles, or twenty-five per cent. of that of the United States, had, in 1879, but 549 miles of railways.[2] To be on a level with the United States in this respect, she should now have 21,829 miles. She needs such public highways more than the United States, for the reason that she has not, like us, a system of navigable rivers to supply transportation facilities from the interior to the seaboard.

The United States had at the time of the last census, in 1870, a total population of 38,558,371.

Mexico had, in 1876, a population of 9,495,157, or twenty-four per cent. of that of the United States six years previous.[3] Her population, then, nearly equals ours in density to the square mile, and from this standpoint also she now requires about twenty thousand miles of railways to be on a level with us.

[1] Poor's Railway Manual for 1880.
[2] Commercial Relations for 1879, vol. i, p. 420.
[3] The Republic of Mexico in 1876, by A. G. Cubas, p. 12.

On the Isthmus itself—an area, as we have previously stated, equal to that of Massachusetts, Connecticut, and Rhode Island combined, there is not a mile of railway except the interoceanic road just commenced. And the states of Oaxaca, Chiapas, Tabasco, and the southern third of Vera Cruz—which states are to a greater or less degree tributary to the Isthmus, and which comprise an area of 66,018 square miles,[1] larger than that of England and Wales—have no existing railway.

Mexico's commercial necessity for railways, and the difficulty of building them except in certain directions, was very clearly stated a few years ago in the Report of the Mexican Committee on Mining Taxes, as follows : "The central table-land of our country is separated from either sea-coast by rugged mountains and deep ravines, breaking it into longitudinal zones of different temperatures and varied productions ; but this fact almost cuts off communication between these zones and the sea-coast east and west. While such natural difficulties exist, increased by territorial extent, manufactures and agriculture cannot thrive, because the cost of transportation is so great we cannot contend with foreign competition, and our vegetable products must be confined to home consumption."[2]

Again, Mexico, unlike the United States, has no trans-continental railway connecting the two oceans; and in most parts of that republic, because of the peculiar natural formation, such a road is either impossible or impracticable.

The Isthmus of Tehuantepec possesses exceptional facilities for supplying this urgent demand for transportation, not only from the interior to the seaboard, but from ocean to ocean. To appreciate its advantages in this respect, a general view of the topographical features of Mexico is necessary.

[1] Article on Mexico, in Johnson's New Universal Cyclopædia.
[2] See Appendix to Blake's " Production of Precious Metals," p. 318.

Topographical Features.

All Mexico, above the Isthmus, is wedge-shaped, and the
Tehuantepec Railway, located across its apex, connects the two
coasts.

Generally speaking, the Cordilleras, table-lands and valleys
all run parallel with the sides of the wedge converging at the
Isthmus. Roads parallel with the drift of this formation are
easy of construction, and at right angles are very difficult, except at the apex, where there is a cross valley, or depression.
The formation is described in Lippincott's Gazetteer as follows :
"Starting from Tehuantepec on the shore of the Pacific, latitude 16° 20′ north, we soon reach the plain of Oajaca, at a
height of five thousand five hundred feet, and thence a wheel-
carriage can roll without difficulty to Santa Fé, in New
Mexico (latitude 36° 12′ north), a distance of about fourteen
hundred miles. Though some parts of the route might fall
below the level of two thousand feet, and others nearly reach
an elevation of eight thousand feet, yet the slopes are so
gradual as to offer no serious impediment to the construction
of an easy road ; but the descent from the table-lands to the sea
is everywhere rugged and abrupt, presenting great difficulties
in the way of carrying goods. In going from the City of Mexico (seven thousand four hundred feet) to Acapulco on the
Pacific, the road attains a height of eight thousand six hundred
feet before the rapid descent commences. The railway between
the Capital and Vera Cruz is regarded as one of the greatest
feats of modern engineering." [1]

Adding to this description that contained in McCulloch's
Geographical Dictionary, the reader will have a more complete
view of the natural formation. It states : "The Cordillera, or
chain of mountains, generally regarded as a portion of the
great chain of the Andes that enters Mexico on the south,
where it borders with Guatemala, diverges, as it proceeds
north, into two great arms, like the upper part of the letter Y,

[1] Article on Mexico in Lippincott's Pronouncing Gazetteer of the World.

4

following the line of the coasts on either side. The most westerly of these chains, or that parallel to the shores of the Pacific Ocean, has some very high summits, and preserves its mountainous character until it joins, on the border of the United States, with the Oregon or Rocky Mountains. The other, or eastern arm of the Cordillera, begins to subside after reaching the twenty-first or twenty-second degree of latitude, and ultimately subsides, about the twenty-sixth or twenty-seventh degree of latitude, into the vast plains of Texas. The whole of the vast tract of country between these two great arms, comprising about three-fifths of the entire surface of the empire, consists of a table-land called the plateau of Anahuac, elevated from six thousand to upwards of eight thousand feet above the level of the sea." [1]

'The map preceding this chapter, the hypsometric features of which we have copied from Ravenstein's map, contained in a London publication, corresponds with the above description.[2] It shows that only at the Isthmus is there a depression of less than one thousand feet in elevation, extending from sea to sea.

It also illustrates the general drift of the table-lands toward the Isthmus.

A Railway Era.

The administration of President Diaz will ever be memorable in the history of Mexico as the beginning of a new era of material development and progress, based upon the construction of rail highways of commerce. The Tehuantepec Inter-Ocean Railroad has the proud distinction of being the first North American Company to consummate, in this way, commercial fellowship between the two sister republics. Since its grant was obtained, in June, 1879, Mexico has awarded two other valuable and comprehensive concessions to citizens of the United States, viz.: to the Mexican Central Company, commonly known as the

[1] Article on Mexico in McCulloch's Geographical Dictionary.
[2] See Map in " Peep at Mexico," by J. L. Geiger.

Boston Company, the right of way from El Paso southwardly along the central table-lands, and from some point on this line another road to the Pacific coast; also to the National Mexican Railway Construction Company, commonly known as the Sullivan Palmer party, the right of way from the City of Mexico to Manzanillo, on the Pacific, and from some point on that line to the Texas border.

These three concessions comprise a system of railways estimated to be about two thousand seven hundred miles in extent. It is evident from these facts that Mexico is endeavoring to make up for much valuable time lost since Cortez commenced the work of internal improvements, early in the sixteenth century.

The question naturally arises, on the part of those unfamiliar with modern Mexico, why has she so long neglected her evident self-interest in this respect, and why have not the restless Anglo-American railway builders long since acquired fortunes in that inviting field? The answer to these questions would require a large volume, containing a review of the frequent revolutions (now ended, we trust) of that magnificent but unfortunate country—a consideration of the war of 1846-7 between the United States and Mexico, the memory of which remained for years an obstacle to the introduction of North American enterprises—a notice of the border troubles arising every now and then during the past ten years, and caused by cattle thieves on the one side and disreputable fillibusters on the other, etc., etc.

Suffice it to say the past is being rapidly forgotten, and the business men of the two neighboring countries see great and mutual profits and advantages, to be derived from an interchange of commerce and friendly services.

But there is one point in the history of this new railway movement which a portion of the people of this country seem to misunderstand, and which it may be well to notice. We refer to the supposed indifference of Mexico to the introduction of railway enterprises by North American companies. A part of the Mexican people and politicians may be so blind to their own material welfare as to desire a Chinese wall to forever remain in the pathway of commercial intercourse, but that is not the wish of modern Mexico, as expressed by repeated official acts in the shape of railway concessions. The facts in this respect have recently been officially compiled and stated by Señor Matias Romero, late Secretary of Finance of that republic. In his report, dated January, 1879, after reviewing the individual contracts awarded to North American companies, he gives the following summary :

"From the preceding data, it appears that from the year 1865 Mexico has granted to companies or citizens of the United States, nineteen concessions of railroads in the republic, in the following order : To the line from Paso del Norte to Guaymas, six times ; to that of the Isthmus of Tehuantepec, nine times ; once to each of the lines from Tuxpam to the Pacific ; from Vera Cruz to Anton Lizardo, to the Isthmus of Tehuantepec, and from San Anton Lizardo to Matamoras Izucar, Cuernavaca, and Acapulco ; from Leon to the frontier of the United States, and from Mexico to the Pacific and to the frontier of the United States ; and, as has already been stated, additional concessions have been made to them for the laying of submarine cables, and for other purposes." [1]

He then reviews the history of railway concessions to companies of other nationalities, and concludes as follows : " The summary, then, of all these important data is that, since the year 1866 to this date, the Mexican government has granted, without counting the contracts made with the governors of states, thirty-three railroad concessions, of which nineteen have been given in favor of citizens of the United States, four in favor of citizens of other nationalities, and the remaining ten in favor of Mexicans." [2] And he adds with much force :

"It seems impossible to present a more eloquent demonstration than that which the preceding facts bring up, to show that so far from the assertion being demonstrable that Mexico looks

---

[1] Report of the Secretary of Finance of Mexico, January, 1879, p. 55.
[2] Ditto, p. 58.

with jealousy or hostility upon the building of railroads which would put her in communication with the United States, or that even, without this circumstance, that the roads should be constructed by North American companies, the truth is that she has, on the contrary, granted them a marked preference up to this time." [1]

The Isthmus as an Outlet.

The population of the table-lands is fully three-fourths, and probably four-fifths, of that of the whole republic." [2]

It is also much more dense in the states a little north of the Isthmus, than in those near the boundary of the United States. This shows the commercial importance of a railway outlet from the interior of Southern Mexico. The Isthmus furnishes such an outlet, not only to one, but to two oceans. It is the natural trunk line for numerous tributary roads.

The survey of Barnard and Williams treats of this point as follows : "We have said that what would conduce to the interest of the owners of the grant would also conduce to that of Mexico herself by connecting railroads which, from the necessity of the case, and in order to supply the wants of commerce, would be constructed. The topography of the contiguous portions of the Mexican Republic exhibits such favorable features that, at a comparatively small outlay, roads can be built which will bring into the lap of the Isthmus the mineral resources and dyes of Oaxaca, the indigo of Guatemala and Chiapas, the sugar and cotton of Vera Cruz, and the tobacco, coffee, and chocolate of Tabasco." [3]

Commodore Shufeldt also has, since his survey in 1871, spoken as follows of the necessity for connecting lines from the four surrounding states : "For the more perfect development

---

[1] Report of the Secretary of Finance of Mexico, January, 1879, p. 58.

[2] Estimated from statistics of the various states on p. 12 of the Republic of Mexico in 1876, by A. G. Cubas.

[3] Barnard and Williams' Survey, p. 143.

of this country the main road would need branches extending through these states to the cities of Vera Cruz, Oajaca, and into Chiapas and the harbor of Tonala. These roads can be constructed with comparative ease and cheapness, from the nature of the ground over which they would pass being nearly level plains." [1]

The general characteristics of these four tributary states are, in products, so similar to the Isthmus, that a detailed statement of their probable patronage of the railway is unnecessary. But Oaxaca, which state is perhaps the most interested in a railway outlet at the Isthmus, possesses some elements of wealth which we have not heretofore described. In a report to the State Department in 1877, the official representative of the United States writes : "This state is rich in iron, lead, copper, gold, and silver, all of which are worked in a limited way. The future prosperity of this state depends on the development of its mineral wealth ; when that time comes, the extent and richness of these mines will give Oaxaca a corresponding interest and importance equal to California and Australia." [2] A few other points relative to this state, and which we have omitted in describing the Isthmus, are mentioned by Robinson, who visited Oaxaca about 1820. He says: "In all the mountainous districts of Oaxaca, and more especially in the spacious valleys which are situated from twenty-five hundred to six thousand feet above the level of the sea, we find a soil and climate at least equal, if not superior, to any on the globe. There is not a single article raised in the temperate zone that would not here find a congenial region. Wheat and all kinds of grain yield a return to the cultivator equal to that of the most fertile parts of Europe. The fruits and vegetables of Oaxaca are unrivalled for luxuriance and delicacy, Peaches, pears, apricots and strawberries are here to be found of a size and flavor superior to those of the south of France ; and the variety and

---

[1] His statement in New York *Herald*, Oct. 22, 1879.
[2] Commercial Relations for 1877, p. 740.

excellence of the grape point out the valleys of Oaxaca as the great future vineyard of New Spain." [1]

The testimony of European writers is equally emphatic. Baron Humboldt, who visited Mexico about the beginning of the present century, and whose work, in four volumes, is universally accepted as a standard authority on that country, says : "The intendancy of Oaxaca is one of the most delightful countries in this part of the globe. The beauty and salubrity of the climate, the fertility of the soil, and the richness and variety of its productions, all minister to the prosperity of the inhabitants ; and this province has, accordingly, from the remotest periods, been the center of an advanced civilization." [2] The area of this state is 27,389 square miles, or larger than the combined areas of Vermont, New Hampshire, and Massachusetts. It has a present population of 661,706 [3] souls, yet not a single mile of railway. Its area combined with the other states which need railways to connect with that projected across the Isthmus, viz : Tabasco, Chiapas, and the southern third of Vera Cruz, is, as we have already stated, larger than that of England and Wales. It is difficult to overestimate the advantages to Mexico which would accrue from railway transportation in these states so rich in natural and undeveloped resources.

But the Isthmus may become an outlet for a still broader area by a connecting railway from the capital of Mexico, which has no steam communication with the Pacific coast of its own country. It would also bring the city of Mexico into direct commercial relations with the Orient. One feature of such communication is worthy of notice, viz.: "Asia has been known in all historical times as the sink of silver." We quote this remark from the Report to Congress in 1877 of the U. S. Monetary Commission, commonly known as the Silver Commission. The Report shows that British India alone, during the forty

[1] Memoirs of the Mexican Revolution, by W. D. Robinson, p. 358.
[2] Humboldt's New Spain, vol. ii, p. 188.
[3] The Republic of Mexico in 1876, by A. G. Cubas, p. 12.

years from 1836 to 1875 inclusive, imported, in excess of its exports, silver to the value of about $1,000,000,000, or $25,000,000 annually. It also states that England alone (which is not a silver-producing country) exported, in 1876, to India and China combined silver to the amount of $45,975,438.[1]

Mexico can easily supply the whole of this steady demand. Her total product of silver alone, from 1492 to 1875 inclusive, a period of three hundred and eighty-four years, was $3,262,370,247.[2]

Her present annual product of silver is about $26,000,000.[3]

Silver is also the chief annual export of Mexico. During the year ending June 30, 1873, of the total exports of that country, amounting to $33,168,609, the precious metals constituted over five-sevenths, or $25,373,673, nearly all of which was silver.[4]

Instead of going by way of England and the Suez Canal to the Orient, these enormous exports should go from Mexico directly across the Pacific ocean.

But silver is but one item in the commercial exchanges which the interior of Mexico should develop with the Orient by means of a railway to the Tehuantepec outlet. We mention it simply as a sample of the commercial possibilities of such a road.

As a Transcontinental Highway.

It seems strange that in this progressive age, and in a North American republic, a country which has a coast line on the Pacific of 4,415 statute miles, and on the Gulf of 1,779 miles[5]— a country, too, rich in nearly all the products of the earth, should have no transcontinental railway. The omission is particularly noticeable when we refer to the map and find that nature has provided a convenient pass, only one hundred and

---

[1] Report of Monetary Commission, Senate Report 703, Forty-fourth Congress, Second Session, pp. 74–76.

[2] The Silver Country, by A. D. Anderson, p. 57.   [3] Ditto, pp. 42–3.

[4] Commercial Relations for 1875, p. 1129.

[5] Distances furnished by Navy Department.

fifty miles long from sea to sea. The road will place the commercial centers of the two coasts of Mexico in easy intercommunication. It will connect her western coast with the West Indies, the Gulf and Atlantic ports of the United States, and with Europe. It will connect her eastern coast with the Orient and the western coasts of South and Central America. In brief, it is a matter of national importance to that republic, and, like our own Union Pacific road, will bind her sections together in commercial ties.

CHAPTER IV.

COMMERCIAL RELATIONS TO THE UNITED STATES.

"The proximity of the Isthmus to our shores, the salubrity of the climate, the adaptness of the ground for the construction of a railroad, and the great diminution of distance in comparison with more southern routes between our Atlantic and Pacific possessions, all conspire to point it out as far preferable to any other route outside of our own territory."

From official letter by LEWIS CASS, *while Secretary of State, in 1857. See Senate Exec. Doc. No. 72, Thirty-fifth Congress, First Session, pp. 39-40.*

CONSIDERED from the standpoint of either coast of the United States, the Atlantic, Pacific, or Gulf, the Isthmus of Tehuantepec is preëminently the most serviceable to our commerce. As Commodore Shufeldt forcibly said in his letter transmitting to the Secretary of the Navy the Report of the Survey of 1871: "Each Isthmus rises into importance as it lies nearer to the center of American political and commercial influence, and the intrinsic value of this eminently national work ought to be based upon the inverse ratio of the distance from that center.[1]

The Mississippi Valley is fast becoming that center, and as the natural relations between its great river system and Tehuantepec are most intimate, we will first consider the question from that standpoint.

To the Mississippi Valley.

Directly opposite Tehuantepec is the sole outlet of the Mississippi River and its forty-two navigable tributaries (forty-three rivers in all), comprising a system of inland navigation through

[1] Shufeldt's Survey, p. 20.

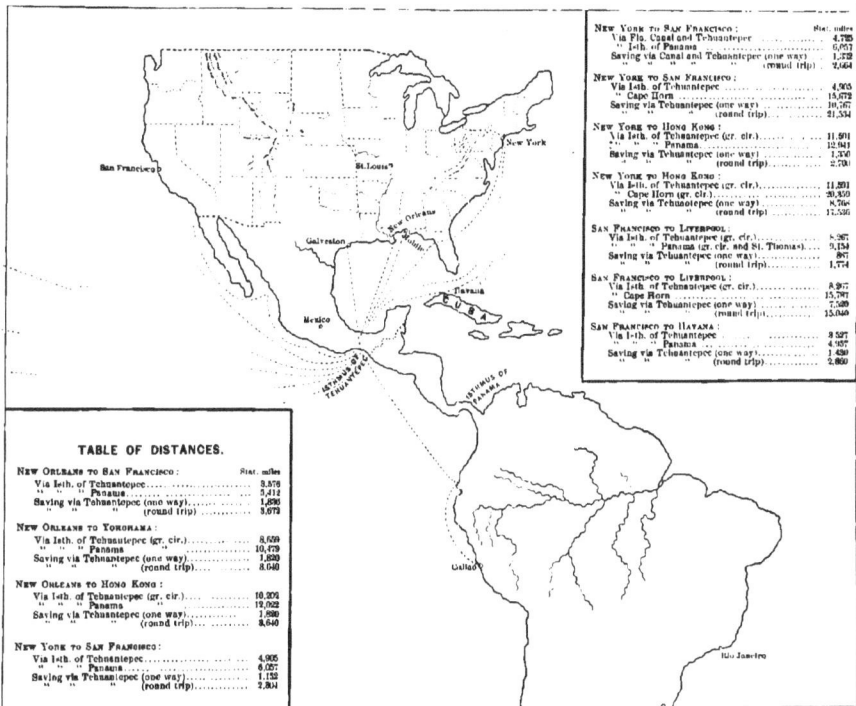

TABLE OF DISTANCES.

NEW ORLEANS TO SAN FRANCISCO: Stat. miles
Via Isth. of Tehuantepec.................... 3,876
"   "   " Panama........................ 5,412
Saving via Tehuantepec (one way)........... 1,836
"   "   (round trip)......... 3,672

NEW ORLEANS TO YOKOHAMA:
Via Isth. of Tehuantepec (gr. cir.)........ 8,659
"   "   " Panama..................... 10,479
Saving via Tehuantepec (one way)........... 1,820
"   "   (round trip)......... 8,640

NEW ORLEANS TO HONG KONG:
Via Isth. of Tehuantepec (gr. cir.)........ 10,202
"   "   " Panama...................... 12,022
Saving via Tehuantepec (one way)........... 1,820
"   "   (round trip)......... 3,640

NEW YORK TO SAN FRANCISCO:
Via Isth. of Tehuantepec................... 4,905
"   "   " Panama...................... 6,057
Saving via Tehuantepec (one way)........... 1,152
"   "   (round trip)......... 2,304

NEW YORK TO SAN FRANCISCO: Stat. miles
Via Flo. Canal and Tehuantepec............. 4,785
"   Isth. of Panama...................... 6,057
Saving via Canal and Tehuantepec (one way).. 1,272
"   "   (round trip)......... 2,464

NEW YORK TO SAN FRANCISCO:
Via Isth. of Tehuantepec................... 4,905
"   Cape Horn........................... 15,672
Saving via Tehuantepec (one way)........... 10,767
"   "   (round trip)......... 21,534

NEW YORK TO HONG KONG:
Via Isth. of Tehuantepec (gr. cir.)........ 11,591
"   "   " Panama..................... 12,941
Saving via Tehuantepec (one way)........... 1,350
"   "   (round trip)......... 2,700

NEW YORK TO HONG KONG:
Via Isth. of Tehuantepec (gr. cir.)........ 11,591
"   Cape Horn (gr. cir.)................ 20,859
Saving via Tehuantepec (one way)........... 8,768
"   "   (round trip)......... 17,536

SAN FRANCISCO TO LIVERPOOL:
Via Isth. of Tehuantepec (gr. cir.)........ 8,267
"   "   " Panama (gr. cir. and St. Thomas).. 9,154
Saving via Tehuantepec (one way)........... 887
"   "   (round trip)......... 1,774

SAN FRANCISCO TO LIVERPOOL:
Via Isth. of Tehuantepec (gr. cir.)........ 8,267
"   Cape Horn........................... 15,787
Saving via Tehuantepec (one way)........... 7,520
"   "   (round trip)......... 15,040

SAN FRANCISCO TO HAVANA:
Via Isth. of Tehuantepec................... 3,527
"   "   " Panama...................... 4,957
Saving via Tehuantepec (one way)........... 1,430
"   "   (round trip)......... 2,860

TEHUANTEPEC-ITS RELATION TO THE UNITED STATES.

the productive center of the United States to the extent of
15,710 miles, as follows : [1]

Rivers.	Heads of Navigation.	Miles Navigable.
1. Mississippi	St. Anthony's Falls, Minn	2,161
2. Missouri	Great Falls, Mon	3,127
3. Ohio	Pittsburg, Pa	1,021
4. Red	State Shoals, Tex	986
5. Arkansas	Wichita, Kan	884
6. White	War Eagle, Mo	779
7. Tennessee	Knoxville, Tenn	759
8. Cumberland	Waitsboro, Ky	609
9. Yellowstone	Belle Butte, Mon	474
10. Ouachita	Camden, Ark	384
11. Wabash	Lafayette, Ind.	365
12. Alleghany	Olean, N. Y	325
13. Osage	Papinville, Mo	303
14. Minnesota	Patterson's Rapids, Minn	295
15. Sunflower	Clarksdale, Miss	271
16. Illinois	La Salle, Ill	270
17. Yazoo	Greenwood, Miss	228
18. Black	Perkins, Mo	212
19. Green	Greensburg, Ky	200
20. St. Francis	Wittsburg, Mo	180
21. Tallahatchie	Hill's Place, Miss	175
22. Wisconsin	Portage City, Wis	160
23. Deer Creek	Stoneville, Miss	116
24. Tensas	Lake Providence, La	112
25. Monongahela	Morgantown, West Va	110
26. Kentucky	Cogar's Landing, Ky	105
27. Bartholomew	Baxter, Ark	100
28. Kanawha	Great Falls, W. Va	94
29. Muskingum	Dresden, O	94
30. Chippewa	Chippewa Falls, Wis	90
31. Iowa	Iowa City, Ia	80
32. Big Hatchie	Bolivar, Tenn	75
33. St. Croix	Falls St. Croix, Wis	65
34. Rock	Sterling, Ill	64
35. Black, La	Mouth Ouachita, La	61
36. Macon	Floyd, La	60
37. Bœuf	Rayville, La	55
38. Big Horn	Ft. Custer, Mon	50
39. Clinch	Clinton, Tenn	50
40. Little Red	Searcy's, Ark	49
41. Cypress and Lake	Jefferson, Tex	44
42. Big Black	Bovina, Miss	35
43. Dauchite	Minden Landing, La	33
Total miles navigable		15,710

[1] The Mississippi and Tributaries, by A. D. Anderson, chap. ii.

The navigable parts of these water-ways intersect or border the following twenty-two States and Territories:

Alabama,	Iowa,	Missouri,	Pennsylvania,
Arkansas,	Kansas,	Montana,	Tennessee,
Dakota,	Kentucky,	Nebraska,	Texas,
Illinois,	Louisiana,	New York,	W. Virginia,
Indiana,	Minnesota,	Ohio,	Wisconsin.
Indian Ty.,	Mississippi,		

Omitting New York, which has but a trifling interest in one tributary ; Pennsylvania, which, although a large patron of the Ohio in shipments from Pittsburgh, is chiefly tributary to the Atlantic coast ; West Virginia, which is also chiefly tributary to the Atlantic; Alabama, whose chief outlet is at the Gulf ; Texas, which, although a large patron of Red River, exports her products mainly from Gulf ports ; and the three Territories, Montana, Dakota, and Indian Territory, whose annual products are not separately stated by the Department of Agriculture, and we have left fourteen States tributary, to a greater or less degree, to the Mississippi River system, having, as we have above remarked, but one outlet, and that near and opposite Tehuantepec. The fourteen States which constitute the Mississippi Valley are, then, substantially as follows :

Arkansas,	Kansas,	Mississippi,	Ohio,
Illinois,	Louisiana,	Missouri,	Tennessee,
Indiana,	Kentucky,	Nebraska,	Wisconsin,
Iowa,	Minnesota,		

By contrasting their great staple products for the year 1879 with those of the whole United States, we find they produced—
Eighty-one per cent of the corn, or 1,263,348,700 bushels ;
Sixty-four per cent of the wheat, or 289,708,550 bushels ;
Sixty per cent of the oats, or 222,028,100 bushels ;
Fifty-five per cent of the tobacco, or 216,842,000 pounds ;
Forty-nine per cent of the cotton, or 2,616,763 bales ; and contained—

Sixty-six per cent, in value, of live hogs, or $97,455,823.[1]

While these water-ways which tap the great and productive Valley have an outlet at the Gulf, the Gulf itself has no commercial outlet on the west, or to the Pacific Ocean. This is a commercial absurdity, as a glance at the map preceding this chapter will illustrate.

The necessity for direct communication between the mouth of the Mississippi and the Pacific is increasing more rapidly than the general public are, perhaps, aware. During the single fiscal year following the completion of the jetties at the mouth of the river, the year ending June 30, 1880, the domestic exports of New Orleans increased $26,456,077 ; the total exports for the year ending June 30, 1879 being $63,782,426, and for the year ending June 30, 1880, $90,238,503.[2] This new commercial movement from the Valley States, down the river to the Gulf, is likely to increase from year to year until it assumes immense proportions. Important railway lines, also, are seeking an outlet at New Orleans.

In brief, the products and manufactures of the Valley are seeking their natural outlet to the seaboard. They should now seek markets in the countries around two oceans instead of one. The designs of nature were never more clearly indicated than that the Mississippi Valley and the countries of the Pacific Ocean should have an interchange of commerce across Tehuantepec. The saving of distance over the present isthmus route *via* Panama will be as follows :[3]

NEW ORLEANS TO SAN FRANCISCO,	Statute Miles.
Via Isthmus of Tehuantepec	3,576
" " Panama	5,412
Saving, *via* Tehuantepec (one way)	1,836
" " (round trip)	3,672

[1] Report of Department of Agriculture for 1879.
[2] From Bureau of Statistics.
[3] Prepared by U. S. Coast Survey. See Appendix.

NEW ORLEANS TO YOKOHAMA,   Statute
                                                     miles.
Via Isthmus of Tehuantepec (great circle)..... 8,659
"   "   Panama   "   " 10,479
Saving, via Tehuantepec (one way) 1,820
"   "   (round trip)......... 3,640

NEW ORLEANS TO HONG KONG,
Via Isthmus of Tehuantepec (great circle)..... 10,202
"   "   Panama   "   " 12,022
Saving, via Tehuantepec (one way)........... 1,820
"   "   (round trip)......... 3,640

NEW ORLEANS TO SIDNEY (AUSTRALIA),
Via Isthmus of Tehuantepec (great circle)..... 9,298
"   "   Panama   "   " 10,451
Saving via Tehuantepec (one way)............ 1,153
"   "   (round trip).......... 2,306

NEW ORLEANS TO MAZATLAN,
Via Isthmus of Tehuantepec................... 2,142
"   "   Panama..... 3,966
Saving, via Tehuantepec (one way)........... 1,824
"   "   (round trip)......... 3,648

To the Valley and Gulf States Combined.

The relations of Tehuantepec to the Gulf States, Texas, Alabama, and Florida, are nearly the same as to the mouth of the Mississippi. The saving of distance from Galveston to the countries of the Pacific is a little greater than from New Orleans; and from Mobile, Key West, and Fernandina, a little less.

Texas has a coast line opposite Tehuantepec of 378 statute miles. Her area is 274,356 square miles, or larger than that of France, England, Wales, and Belgium combined. In 1878, on $1\frac{3}{10}$ per cent of her area, she produced 497,310,000 pounds, or 1,105,133 bales of cotton,[1] being the largest cotton crop of any State.

[1] See Report of Department of Agriculture for 1878.

The total foreign commerce, exports and imports combined, of her Gulf ports was, for the year ending June 30, 1879, $22,682,334 in value.[1]

The total foreign commerce of Mobile, during the same year, was $6,764,446 in value.[2]

The necessity for a commercial outlet from the Gulf to the Pacific becomes still more apparent when we combine the leading products of the Valley with those of the remaining three Gulf States ; in other words, the products of the above-mentioned fourteen river States, with those of Texas, Alabama, and Florida.

Their combined products for the year 1879, contrasted with those of the whole United States for the same year, shows that they yielded in quantity—

Eighty-five per cent of the corn ;

Sixty-five per cent of the wheat ;

Sixty-two per cent of the oats ;

Seventy-eight per cent of the cotton ;

Fifty-five per cent of the tobacco, and contained

Seventy-two per cent in value of live hogs.

Yet there is no commercial highway from the Gulf to the Pacific, either canal or railway. Such an omission is a discredit to American enterprise.

To the Atlantic States.

The voyage from the chief port of the Atlantic States, *via* Tehuantepec to San Francisco, will be along our own coast and that of a friendly republic, with the following saving of distance over Panama :[3]

NEW YORK TO SAN FRANCISCO,	Statute miles.
Via Isthmus of Tehuantepec	4,905
" " " Panama	6,057
Saving *via* Tehuantepec (one way)	1,152
" " " (round trip)	2,304

[1] Commerce and Navigation for 1879.   [2] Ditto.
[3] Prepared by U. S. Coast Survey. See Appendix. [4]

NEW YORK TO SAN FRANCISCO,   Statute miles.
   Via proposed Florida Canal and Tehuantepec.. 4,725
   " Isthmus of Panama..................... 6,057
   Saving *via* Fla. Canal & Tehuantepec (one way) 1,332
    "   "   "   " (round trip) 2,664

NEW YORK TO SAN FRANCISCO,
   Via Isthmus of Tehuantepec................. 4,905
   " Cape Horn 15,672
   Saving *via* Tehuantepec (one way) 10,767
    "   "   " (round trip) 21,534

It will be observed, by a reference to the map, that the proposed Florida Canal has no connection with the Isthmus of Panama, but a very intimate one with Tehuantepec and the Mississippi River. The able official report of Gen. Q. A. Gilmore, showing its commercial necessity, was recently submitted to Congress, and is now awaiting their action.

From New York to the Orient, the comparative distances are as follows : [1]

NEW YORK TO YOKOHAMA,   Statute miles.
   Via Isthmus of Tehuantepec (great circle)..... 9,996
   "   "   " Panama   "   " 11,013
   Saving *via* Tehuantepec (one way)............ 1,017
    "   "   " (round trip).......... 2,034

NEW YORK TO YOKOHAMA,
   Via Isthmus of Tehuantepec (great circle)..... 9,996
   " Cape Horn   "   " 19,783
   Saving *via* Tehuantepec (one way)............ 9,787
    "   "   " (round trip)..........19,574

NEW YORK TO HONG KONG,
   Via Isthmus of Tehuantepec (great circle).....11,591
   "   "   " Panama   "   " 12,941
   Saving *via* Tehuantepec (one way)............ 1,350
    "   "   " (round trip)........ 2,700

[1] Prepared by U. S. Coast Survey. See Appendix.

NEW YORK TO HONG KONG,   Statute
                                                       miles.
  Via Isthmus of Tehuantepec (great circle).....11,591
  " Cape Horn   "   " 20,359
Saving *via* Tehuantepec (one way)............ 8,768
  "   "   "   (round trip)..........17,536

While the distances from Portland, Boston, Philadelphia, Baltimore, and Norfolk to San Francisco and the Orient will differ from the above figures, the saving will be substantially the same as from the New York standpoint.

To the Pacific States.

At the close of our late civil war the United States loaned to the Union and Central Pacific Railroads $53,121,632 in bonds, and granted 20,000,000 acres of public lands to help build those roads, and thereby strengthen the political and commercial ties between the Pacific States and the rest of the Union. The Tehuantepec Railroad will add other ties of great strength, and at no expense to the general government. The saving of distance over Panama we have just stated from the New Orleans and New York standpoints, and need not therefore repeat the tables in this connection.

But there will be *via* Tehuantepec a saving of distance in other directions of great commercial importance to the Pacific ports. We refer particularly to direct trade between San Francisco and Liverpool and Havana. The distances are as follows : [1]

SAN FRANCISCO TO LIVERPOOL,   Statute
                                                      miles.
  Via Isth. of Tehuantepec (great circle)........ 8,267
  " " " Panama   " and St. Thomas. 9,154
Saving *via* Tehuantepec (one way)............ 887
  "   "   "   (round trip).......... 1,774

SAN FRANCISCO TO HAVANA,
  Via Isthmus of Tehuantepec................. 3,527
  "   "   " Panama..................... 4,957
Saving *via* Tehuantepec (one way)............ 1,430
  "   "   "   (round trip)......... 2,860

[1] Prepared by U. S. Coast Survey. See Appendix.

The great interest California has in a short route to Liverpool *via* the Isthmus of Tehuantepec may be seen from a glance at the statistics of her wheat product.[1] In 1878 she produced 41,990,000 bushels, the yield being seventeen to the acre. In 1879 she produced 35,000,000 bushels, the yield per acre being the same.

In 1880, according to the preliminary estimate of the Department of Agriculture, she has under cultivation 3,043,000 acres of wheat. If the yield per acre proves to be as great as in 1878 and 1879, her product for the present year will be 51,731,000 bushels.

A large portion of this enormous annual product goes to the Liverpool market. It can save by way of Tehuantepec, in contrast with Panama, as we have above stated, 1,774 miles on the round trip.

But all, or nearly all, of these grain shipments were made by way of Cape Horn. The remarkable saving of distance *via* Tehuantepec over the Cape Horn route may be seen from the following contrast.[2]

San Francisco to Liverpool,	Statute miles.
Via Isthmus of Tehuantepec (great circle)....	8,267
" Cape Horn.............................	15,787
Saving *via* Tehuantepec (one way)...........	7,520
" " (round trip)........	15,040

The total annual commerce, exports and imports combined, of the various West India Islands is, in value, $309,769,000,[3] in which grand total the port of San Francisco has no recorded direct participation. As the Tehuantepec Railroad will shorten the distance between the chief port of the Pacific States and Havana, the chief port of those fertile islands, 2,860 statute miles on the round trip, it is to be hoped that an extensive and profitable interchange of commodities will be the result.

---

[1] Compiled from official statistics of U. S. Department of Agriculture.
[2] See distances in Appendix of this paper.
[3] Commercial Relations of U. S. for 1879, p. 32.

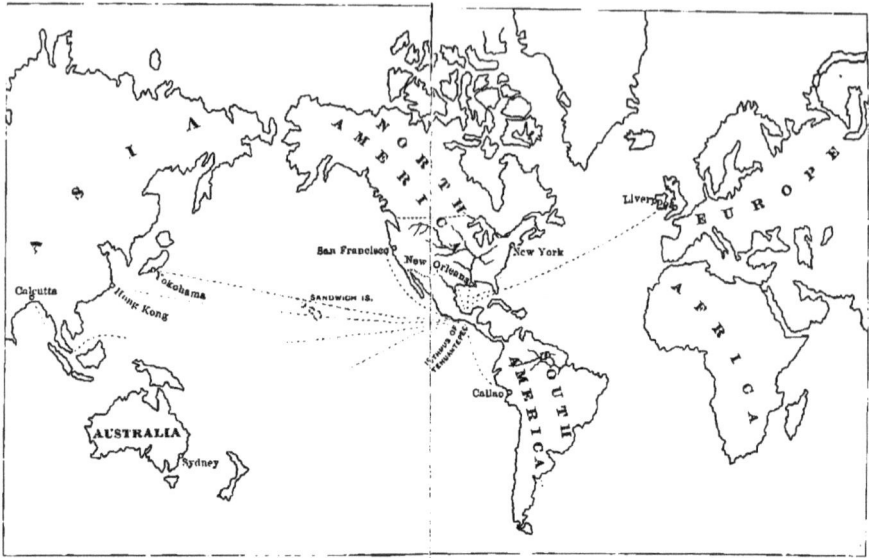

TEHUANTEPEC-ITS RELATION TO THE WORLD.

CHAPTER V.

Its Central Position.

A GLANCE at a map of the world shows that the Isthmus of Tehuantepec is located midway between North and South America ; the two coasts of America ; Europe and Eastern Asia ; Europe and Australia; and on the ocean highway between Europe and the western coast of America ; New York and the Orient ; New York and Australia ; the Mississippi Valley and the Orient ; the Mississippi Valley and Australia ; the Mississippi Valley and the western coast of America ; and the West Indies and the countries of the Pacific Ocean. In brief, its natural relations to the commercial nations are more general and advantageous than those of any other Isthmus of the earth.

Possible Present Patronage.

The recent report on the Inter-Oceanic Canal question by the United States Bureau of Statistics, gives an estimate, from the Panama standpoint, of the possible commerce across the American Isthmus during the last attainable year, as follows :

" Number of Vessels and Amount of Tonnage which might have passed through the proposed canal if it had been constructed.

[N. B.—This table is based upon statistics of the latest year for which the requisite data can be obtained.]

	No. of Vessels.	Tons.
1. Average number of vessels and amount of tonage entered at and cleared from either side of the Isthmus of Panama, annually, in trade with all nations.[1] (Appendix 2 and 3)........	338	533,000
2. Vessels entered at and cleared from Pacific ports of the United States in trade around Cape Horn with Atlantic ports of the United States, during the year ended June 30, 1879.[2] (Appendix 4)	75	120,662
3. Vessels entered at and cleared from Atlantic ports of the United States in trade with foreign countries west of Cape Horn, during the year ended June 30, 1879. (Appendix 5)	273	247,567
4. Vessels entered at and cleared from Pacific ports of the United States in trade with foreign countries east of Cape Horn, during the year ended June 30, 1879. (Appendix 6)	455	551,929
5. Vessels entered at and cleared from ports of the several countries of Europe in trade around Cape Horn with foreign countries other than the United States, during the latest year for which the data can be stated with respect to each country. (Appendix 7)................	1,644	1,462,897
6. Vessels entered at and cleared from ports of British Columbia in trade with countries of Europe, during the year ended June 30, 1879 ..	33	22,331
Total	2,818	2,938,386

The foregoing statement, as already mentioned, is based upon the single condition of distance, and embraces only shipping when employed upon voyages which are shorter by the way of the proposed canal than by any other practicable route."

[1] An estimate from the report of the United States Consul at Panama (Appendix 2), and from a statement compiled from British consular returns (Appendix 3).
[2] Compiled from special reports by collectors of customs.

This estimate has recently been reviewed by the Executive Committee of the proposed Nicaragua Ship Canal Society, corrected and changed so as to represent the "*probable* Canal trade for the year 1880." Their conclusion is as follows :

"Item 1	....	746,200 tons
" 2	....	250,000 "
" 3	....	230,447 "
" 4	....	1,500,000 "
" 5	....	957,448 "
" 6	....	22,331 "
	Total	3,706,426 tons."

We will not attempt to review either of these estimates, unless it be to substitute in the latter the word "possible" for "probable."

We assume that what is possible for an Inter-oceanic Canal is substantially so for a Railroad.

Possible Future Patronage.

The Chief of the Bureau of Statistics confines his estimate almost exclusively to the commerce which "*might have*" passed through the Isthmus during the last attainable year. He seems to have forgotten that he was writing in progressive America, where people live more in the *future* than the *past*, where material development progresses with lightning-like rapidity, and industrial and commercial statistics double every few years.

Again, he makes his calculations from a standpoint which we must discard in forming an estimate of the future, for the Isthmus of Panama has several disadvantageous features. It is too remote from the United States. The commercial route by way of it is, as we have shown in a previous chapter, far greater between our Gulf and Atlantic ports, Liverpool, Havana, etc., on the one side, and the principal countries of the Pacific, on the other, than by way of Tehuantepec.

The states and countries which immediately surround it have less commercial possibilities than those around Tehuantepec. A highway across it cannot possibly open up so many new markets as can the latter route.

Its surroundings in the Pacific are particularly disadvantageous. The Report of Commander Selfridge on the Darien Ship Canal says : "Lying near the equator, but generally a little north of it, is a belt, some four or five degrees wide, of calms, rains, and light baffling winds, that separates the wind systems of the north Pacific from those of the south. Its average northern limit may be placed at 8° north and its southern at 3° north, but both are very variable. This is often spoken of as a 'calm belt,' which term is calculated to mislead one as to the nature of the weather to be expected within these limits. Neither does the term 'region of variable winds' appear satisfactory, as it does not express the peculiar character of the weather. '*Doldrums*' seems to be the correct word, for, although it may be, as some say, 'uncouth,' it is the only single word that conveys to the mind of the seaman all that can be expressed by saying 'a region of calms, squalls, light baffling winds, and storms of wind and rain.'"[1] On this same subject Capt. DeKraft, Hydrographer of the United States Bureau of Navigation, has recently stated as follows :

" In order to reach the region of the trade-winds, sailing vessels from Panama, bound to India, Japan, California, or the northwest coast, in coming out of the bay and afterward making the necessary northing, will be obliged to make about six hundred miles through variable winds and vexatious calms before finding themselves in as good a position to make their westing as the vessels leaving the coast of Nicaragua."[2]

That eminent authority on the physical geography of the sea, Lieut. M. F. Maury, says, after describing the winds and currents around the American Isthmus : " You will observe at a

---

[1] House Mis. Doc. No. 113, Forty-second Congress, Third Session, p. 231.
[2] From letter in N. Y. *Herald*, September 29, 1880.

glance that the Isthmus of Panama, or Darien, is, on account of these winds and calms, in a purely commercial point of view, the most out of the way place of any part of the Pacific coast of inter-tropical America."[1]

These vexatious *doldrums* do not extend so far north as the Isthmus of Tehuantepec. For this and the other reasons before stated, Tehuantepec is a more desirable standpoint from which to estimate the possible future patronage.

FROM THE NATURAL GROWTH OF COMMERCE.—As it is from American commerce that the railroad will derive its chief support, it is important to consider its rapid growth. During the fiscal year ending June 30, 1869, the total foreign trade of the United States (exports and imports of merchandise) was $703,624,076 in value. Since then it has, through a natural growth, more than doubled, being for the fiscal year ending June 30, 1880, $1,503,586,897.

The products of corn and wheat of the fourteen States of the Mississippi Valley, which have through its great river system so close a connection with Tehuantepec, have changed as follows during the past ten years, viz.:[2]

CORN.

For year 1869.............	639,850,000	bushels
" " 1879.............	1,263,348,700	"
Increase.............	623,498,700	"

WHEAT.

For year 1869.............	161,737,000	bushels
" " 1879.............	289,708,550	"
Increase.............	127,971,550	"

Texas, also, whose outlets at the seaboard are near Tehuantepec, shows a remarkably rapid development. She had under

[1] See statement by Lieut. Collins, of U. S. N., before House Committee on Inter-oceanic Canals, February 28, 1880, p. 41 of testimony.

[2] Compiled from Official Reports of U. S. Department of Agriculture.

cultivation, in 1870, 104,700 acres of wheat, which acreage has
increased to 459,000 in 1880.[1] This product in bushels was as
follows, from 1872 to 1878 :[2]

1872 ....1,377,000 bushels
1873 ....1,404,000 "
1874 ....1,474,000 "
1875 ....2,510,000 "
1876 ....4,750,000 "
1877 ....4,800,000 "
1878 ....7,200,000 "

Her cotton product increased from 350,628 bales in 1870, to
1,105,133 bales in 1878.[3] The product for this year bids fair
to be still larger.

On the other side of the Isthmus we find as remarkable an
increase in the wheat product of California, the surplus portion of which now passes the doorway of Tehuantepec in going
around Cape Horn to the Liverpool market. The following is
her product for the last nine years :[4]

1871 .... 16,757,000 bushels
1872 .... 25,600,000 "
1873 .... 21,504,000 "
1874 .... 28,380,000 "
1875 .... 23,800,000 "
1876 .... 30,000,000 "
1877 ...●.... .... 22,000,000 "
1878 .... 41,990,000 "
1879 .... 35,000,000 "

She has under wheat cultivation, the present year, 3,043,000
acres,[5] and if the yield per acre is as great as in 1878 and 1879,
the record of her product for 1880 will be 51,731,000 bushels.

[1] From statistics of Department of Agriculture—that for 1880 being a
preliminary estimate.
[2] Ditto.   [3] Ditto, and United States Census.
[4] From Reports of Department of Agriculture.
[5] Preliminary estimate of Department of Agriculture.

The foreign commerce of Australia, which, by means of a railway across Tehuantepec, will have direct connection with that of the Mississippi Valley and our Atlantic ports, has increased in value from $259,625,097 in 1868, to $374,282,078 in 1878.[1] These figures represent her total commerce, exports and imports of merchandise, coin, and bullion.

The total foreign trade of New Zealand, which island bears a similar commercial relation to Tehuantepec, has increased in value from $45,568,338 in 1868 to $71,493,396 in 1878.[2] These figures also represent a combination of exports and imports of merchandise, coin, and bullion.

The above are but a portion of the statistics which we might exhibit to illustrate the remarkable growth of the products and commerce of the various countries surrounding and interested in the Tehuantepec route. As they are a record of a natural and not exceptional growth, they are a safe guide for estimating the future regular growth. We conclude, then, that the products and trade of those countries will double again in a few years, and with them the patronage of the Isthmus Railway.

FROM EXCEPTIONAL CAUSES.—The Gulf of Mexico, or "American Mediterranean," with an area of about 700,000 square miles, or nearly as large as the Mediterranean of Europe, has never had a commercial outlet to the Pacific Ocean, either canal or railway. It is difficult to over-estimate the amount of new trade which the proposed Tehuantepec outlet will create. It is safe to say it will be immense, for the Gulf is surrounded by countries rich in natural resources, and great in commercial possibilities.

The Mississippi River has never, until recently, had an unobstructed outlet at the Gulf. The recent improvement, by jetties, at its mouth, has given a wonderful impetus to shipments down the river. Since their completion, a little more

---

[1] Compiled from Statistical Abstract No. 16 of United Kingdom.
[2] Ditto.

than a year ago, the exports of grain and cotton from New
Orleans have increased remarkably, as may be seen from a contrast with those of the year previous, viz. : [1]

Year ending Aug. 31.	WHEAT.		
1879	1,868,084	bushels	
1880	5,344,510	"	
Increase in one year	3,476,426	"	
	CORN.		
1879	5,117,825	bushels	
1880	9,863,790	"	
Increase in one year	4,745,965	"	
	COTTON.		
1879	1,435,336	bales	
1880	1,696,718	"	
Increase in one year	261,382		

As this new commercial movement, southward, has its outlet
at the Gulf, near and directly opposite Tehuantepec, it is
apparent that its future patronage of the supplemental Isthmus outlet to the Pacific will develop very rapidly.

The important port of Havana has never had a direct commercial highway to California, China, and Japan. With a short
connection across Tehuantepec, new markets will be the natural
result.

The Railroad will greatly facilitate an interchange of trade
between the Mississippi Valley and the western coast of South
'America, for vessels in sailing from the west coast of Tehuantepec can steer clear of the *doldrums* to which they are now sub-
jected in sailing from Panama.

China has recently made an immense gap in her conservative
wall by sending, for the first time, and under her own flag, a
steamer to one of our ports.[2] Direct trade relations between
that country and the Mississippi Valley by way of Tehuantepec
is only a question of time.

[1] From New Orleans Price Current, September 1, 1880.
[2] See N. Y. *Herald*, August 14 and 15, 1880.

Mexico has never had a transcontinental railway connecting her two coasts—her western coast with foreign countries on the Atlantic, or her eastern coasts with foreign countries on the Pacific Ocean. This is perhaps the chief exceptional reason for expecting a remarkable future patronage of the Tehuantepec route. The new era of development upon which Mexico has just entered, is in rapidity, riches, and brilliancy likely to astonish the civilized world—for that republic is one magnificent mine and plantation combined, unstocked and undeveloped.

Gen. Grant, ever since his recent visit to Mexico, has been trying to impress upon the business men of the United States the importance of railway and commercial connection between the two Republics. In a recent speech before the Mechanics' Association of Boston, he said : "There is no secret of the capacity of Mexico for production, if you think about it. We now do an importation business of nearly $200,000,000 of tropical and semi-tropical products. Mexico could produce the whole of them, if she had railroads to give her an outlet for them, and her people have the industry to do it—a fact not generally credited, nor did I believe it until I was there this last time."[1]

On the subject of Mexican possibilities, Señor Romero, while Secretary of Finance, in the early part of last year, wrote that, if the mines were worked as in California and Nevada, "it can be assured that without any great effort it would be easy to augment the annual production of silver from about twenty-five millions—about the average it has amounted to for some years past—to a hundred or a hundred and fifty millions."[2]

This new development, already commenced in a wholesale manner by railway grants to American companies to the extent of about two thousand seven hundred miles, will have a decided effect in increasing the regular patronage of the Inter-ocean Railroad.

Again, there is at the present time, aside from the regular

[1] N. Y. *Tribune*, October 14, 1880.
[2] Report of Secretary of Finance of Mexico, p. 128.

growth of our foreign commerce, a special and popular demand
throughout the United States for new markets and a new merchant marine. The consummation of such national enterprises
will have an immediate and marked effect upon the business of
the Tehuantepec route.

Having stated the leading facts bearing upon the *possible*
present and future patronage of the Railroad, we prefer that the
reader should draw his own conclusion about the *probable* tonnage and value.

CHAPTER VI.

IT is fortunate that this Isthmus, so near our shores and so advantageous to our coastwise and foreign trade, is located in a sister republic.

The two countries are now rivaling each other in the arts of peace, and seeking closer commercial intercourse.

They furnish in their history many striking political parallels. Both were at an early date occupied by European civilization, the one by Spain, in the sixteenth century, and the other by England, in the seventeenth. Both developed in strength, sought and struggled for independence. The one threw off the European yoke in 1776, the other in 1821.

They adopted similar forms of government and constitutions, and, what is still more important for the purposes of this paper, they have a mutual appreciation of and respect for the Monroe doctrine. When Maximilian had for the time being overthrown the young republic, that historic doctrine was reaffirmed by the United States, and became largely instrumental in undermining the foreign and monarchical political structure which Napoleon III had sought to build up in the New World. The structure soon tottered and fell, and Mexico naturally respects the political doctrine which helped regain her liberties. For these reasons political complications concerning the interoceanic highway across this Isthmus are unlikely to arise between the two Republics, or between the United States and European powers.

To quarrel there must be two sides, and as neither France nor England possesses a grant of the right of way across Tehuantepec, nor is likely to have one, either by original concession or by assignment, there can, here, be no basis for an international dispute over the Monroe doctrine.

Mexico's friendly desire to coöperate with the United States in opening the Isthmus to commerce is well illustrated by her repeated concessions to American citizens, and by the treaty of 1853, commonly known as the Gadsden Treaty. In it she stipulates as follows :

"The Mexican government, having on the 5th of February, 1853, authorized the early construction of a plank and rail road across the Isthmus of Tehuantepec, and to secure the stable benefits of said transit way to the persons and merchandise of the citizens of Mexico and the United States, it is stipulated that neither government will interpose any obstacle to the transit of persons and merchandise of both nations ; and at no time shall higher charges be made on the transit of persons and property of citizens of the United States than may be made on the persons and property of other foreign nations, nor shall any interest in said transit way, nor in the proceeds thereof, be transferred to any foreign government. The United States, by its agents, shall have the right to transport across the Isthmus, in closed bags, the mails of the United States not intended for distribution along the line of communication ; also the effects of the United States government and its citizens which may be intended for transit and not for distribution on the Isthmus, free of custom-house or other charges by the Mexican government. Neither passports nor letters of security will be required of persons crossing the Isthmus and not remaining in the country." [1]

As was shown in the Historical Notes, Mexico has "fifteen times" made or renewed to American citizens grants of the right of way across the Isthmus for purposes of a canal or railway. No grant has ever been made to Europeans, and but two assignments to them recognized.

[1] U. S. Statutes, vol. x, pp. 1031–1037.

These facts are ample evidence of Mexico's friendly regard for citizens of the United States, and her willing co-operation in the opening of a commercial highway across her territory.

If for any unforeseen reason a war should occur between either Mexico or the United States on the one side, and some European power on the other, Tehuantepec has peculiar advantages for purposes of defence.

The Gulf of Mexico is nearly surrounded by the two Republics interested in its protection. Its coast-line from Cape Sable, Florida, around to Cape Catoche, Yucatan, is two thousand six hundred and ninety-two miles. It has but one entrance from the ocean, and that is between the extremities of Florida and Yucatan, a distance of only five hundred and fifty-three statute miles, and part of that is closed by the Island of Cuba, leaving for vessels two channels, one of only one hundred and forty-three miles in width, and the other still less, or one hundred and thirty-eight miles.[1]

The question of protection from possible European interference is, then, practically settled by nature.

The political as well as material welfare of Mexico depends very largely upon a system of internal improvements and new commercial relations with the outside world. We have a political as well as business interest in the success and perpetuity of Republican institutions, particularly those of a great and neighboring American nation, whose organic law is so similar to our own. We cannot afford to see Mexico decline. We should rejoice at her welfare and progress. Our political sentiments toward her should be as liberal and friendly as those of France toward us during our Revolutionary struggles. All fillibustering schemes for conquest or plunder inaugurated by a few adventurers on the border should receive from us an emphatic condemnation. The true sentiment for the United States to harbor is well illustrated by the remark of our great and grand President Lincoln, who said, while lamenting the temporary downfall of that Republic at the time of

---

[1] See Table of Distances in Appendix.

Maximilian's conquest, "The Republic of Mexico must rise again." [1]

Setting aside all considerations of pecuniary profit to the Railway Company, or to citizens of the United States and Mexico, the Tehuantepec Inter-Ocean Railroad is of great political importance as a pioneer movement in the direction of closer intercourse and good-fellowship between two great and adjoining North American Republics.

[1] Mexico and the United States, by G. D. Abbott, p. 391.

APPENDIX.

TABLE OF DISTANCES.

U. S. Coast and Geodetic Survey Office,
Washington, *October* 18, 1880.

Mr. A. D. Anderson, Washington, D. C.

Dear Sir: The following table of distances has been prepared by the Hydrographic Inspector in reply to your request of October 1st.

Distances in Statute Miles along the shortest Steamship Routes.

New Orleans to San Francisco: Statute miles.

    Via Isthmus of Tehuantepec............ 3,576

    "   "   Panama.................... 5,412

    Saving *via* Tehuantepec (one way) 1,836

    "   "   (round trip) 3,672

New Orleans to Yokohama :

    Via Isthmus of Tehuantepec (great circle)..... 8,659

    "   "   Panama   "   " 10,479

    Saving *via* Tehuantepec (one way) 1,820

    "   "   (round trip) 3,640

New Orleans to Hong Kong :

    Via Isthmus of Tehuantepec (great circle).....10,202

    "   "   Panama   "   " 12,022

    Saving *via* Tehuantepec (one way) 1,820

    "   "   (round trip) 3,640

NEW ORLEANS TO SIDNEY :   Statute
                                                         miles.
Via Isthmus of Tehuantepec (great circle)..... 9,298
"   "   Panama   "   " 10,451
Saving *via* Tehuantepec (one way)........... 1,153
"   "   (round trip).......... 2,306

NEW ORLEANS TO MAZATLAN,
Via Isthmus of Tehuantepec................... 2,142
"   "   Panama..................... 3,966
Saving *via* Tehuantepec (one way) 1,824
"   "   (round trip)......... 3,648

NEW YORK TO SAN FRANCISCO :
Via Isthmus of Tehuantepec................. 4,905
"   "   Panama..................... 6,057
Saving *via* Tehuantepec (one way) 1,152
"   "   (round trip) 2,304

NEW YORK TO SAN FRANCISCO :
Via proposed Florida Canal & Tehuantepec.... 4,725
"   Isthmus of Panama.................... 6,057
Saving *via* Canal & Tehuantepec (one way).... 1,332
"   "   (round trip).... 2,664

NEW YORK TO SAN FRANCISCO :
Via Isthmus of Tehuantepec 4,905
"   Cape Horn 15,672
Saving *via* Tehuantepec (one way) 10,767
"   "   (round trip)........21,534

NEW YORK TO YOKOHOMA :
Via Isthmus of Tehuantepec (great circle)..... 9,996
"   "   Panama   " 11,013
Saving *via* Tehuantepec (one way)............ 1,017
"   "   (round trip) 2,034

NEW YORK TO YOKOHOMA :

Statute miles.

Via Isthmus of Tehuantepec (great circle) 9,996

" Cape Horn............ " 19,783

Saving *via* Tehuantepec (one way) 9,787

" " (round trip) 19,574

NEW YORK TO HONG KONG :

Via Isthmus of Tehuantepec (great circle) 11,591

" " Panama " 12,941

Saving *via* Tehuantepec (one way) 1,350

" " (round trip) 2,700

NEW YORK TO HONG KONG :

Via Isthmus of Tehuantepec (great circle) 11,591

" Cape Horn............ " 20,359

Saving, *via* Tehuantepec (one way) 8,768

" " (round trip) 17,536

NEW YORK TO SIDNEY :

Via Isthmus of Tehuantepec (great circle) 10,688

" " Panama.... " 11,133

Saving *via* Tehuantepec (one way) 445

" " (round trip) 890

NEW YORK TO SIDNEY

Via Isthmus of Tehuantepec (great circle) 10,688

" Cape Horn............ " 14,739

Saving *via* Tehuantepec (one way)............ 4,051

" " (round trip) 8,102

SAN FRANCISCO TO LIVERPOOL :

Via Isthmus of Tehuantepec (great circle) 8,267

" " Panama ... (" & St. Thomas) 9,154

Saving *via* Tehuantepec (one way) 887

" " (round trip) 1,774

SAN FRANCISCO TO LIVERPOOL :   Statute miles.

 Via Isthmus of Tehuantepec (great circle)..... 8,267

  " Cape Horn 15,787

 Saving *via* Tehuantepec (one way)........... 7,520

  "   " (round trip).........15,040

SAN FRANCISCO TO HAVANA :

 Via Isthmus of Tehuantepec................. 3,527

  " " Panama..................... 4,957

 Saving *via* Tehuantepec (one way)........... 1,430

  "   " (round trip) 2,860

NEW YORK TO ISTHMUS OF TEHUANTEPEC (mouth of Coatzacoalcos River) 2,241

NEW YORK TO ISTHMUS OF PANAMA (port of Aspinwall)... 2,268

SAN FRANCISCO TO ISTHMUS OF TEHUANTEPEC (entrance to Lagoon) 2,495

SAN FRANCISCO TO ISTHMUS OF PANAMA (port of Panama)..3,754

COAST LINE OF GULF OF MEXICO : From Cape Sable, Florida (not including indentations) around to Cape Catoche, Yucatan...............................2,692

ENTRANCE TO GULF OF MEXICO : From Cape Sable to Cuba (Cardinas)....................................... 143

From Cardinas to Cape San Antonio (coast line of northern part of Cuba, not including indentations).. 272

From Cuba (Cape San Antonio) to Cape Catoche...... 138

 Total from Cape Sable to Yucatan 553

Yours respectfully,

C. P. PATTERSON, Sup't.

LIST OF AUTHORITIES.

SURVEYS.

BARNARD, J. G., AND WILLIAMS, J. J.—The Isthmus of Te-huantepec being the result of a survey for a railroad to con-nect the Atlantic and Pacific Oceans made by the Scientific Commission under the direction of Major J. G. Barnard, U. S. Engineer, with a résumé of the climate, local geography, productive industry, fauna and flora of that region. Illus-trated with numerous maps and engravings. Arranged and prepared for the Tehuantepec Railroad Company of New York, by J. J. Williams, principal assistant engineer. New York, 1852 : D. Appleton & Co.

CRAMER, DON AUGUSTIN.—He was a civil engineer, and in 1774, by order of the Viceroy, made a voyage of discovery. He surveyed and reported upon the Isthmus of Tehuantepec. A brief quotation from his Report may be found in English in Garay's publication mentioned on a following page.

MORO, SIGNOR GAETANO.—He was the engineer in charge of the Scientific Commission appointed by Garay, which surveyed the Isthmus in 1842. The substance of his report is given in English in Garay's publication hereafter mentioned.

ORBEGOZO, JUAN DE, AND ORTIZ, DON TADEO DE.—These engi-neers constituted a commission appointed in 1824, the former by the Government of Mexico, and the latter by the State of Vera Cruz, to survey the Isthmus. Each made a report, and extracts from each are given in English in Garay's publication.

SHUFELDT, ROBERT W.—Reports of Explorations and Surveys to ascertain the practicability of a Ship-canal between the Atlantic and Pacific Oceans by way of the Isthmus of Tehuantepec, by Robert W. Shufeldt, Captain U. S. Navy. Made under the direction of the Secretary of the Navy. Senate Ex. Doc. No. 6, Forty-second Congress, Second Session. Washington, 1872 : Government Printing Office.

SIDELL, W. H.—He made a survey of the Isthmus in 1859, as Chief Engineer of the Louisiana Tehuantepec Company. If his report was ever printed, it is now very difficult to find.

POPULAR PUBLICATIONS.

ABBOTT, GORHAM D.—Mexico and the United States. New York, 1869 : G. P. Putnam & Son. It treats largely of interoceanic transit, expressing a preference for Tehuantepec.

BALDWIN, J. D.—Ancient America. New York, 1872: Harper & Brothers. Pages 104–111 are devoted to the ruins of Palenque, near Tehuantepec.

BANCROFT, H. H.—The Native Races of the Pacific States of North America, five volumes. New York, 1876: D. Appleton & Company. Pages 286–423 of Vol. IV relate to the antiquities around Tehuantepec in the States of Oaxaca, Chiapas, and Tabasco, and contain numerous illustrations.

BRASSEUR DE BOURBOURG.—Voyage sur l'Isthme de Tehuantepec. Paris, 1861.

CABRERA, P. F.—Description of the ruins of an ancient city discovered near Palenque in the kingdom of Guatemala in Spanish America. Translated from the original manuscript report of Captain Don Antonio Del Rio, etc., etc. London, 1822 : Henry Berthoud.

CATHERWOOD, F.—Views of Ancient Monuments in Central America, Chiapas, and Yucatan. New York, 1844 : Bartlett & Welford. Plates VI and VII are elegant views of the ruins of Palenque near the Isthmus.

CHALONER AND FLEMING.—The Mahogany Tree. Its botanical characters, qualities, and uses, with practical suggestions for selecting and cutting it in the regions of its growth in the West Indies and Central America ; with notices of the projected interoceanic communications of Panama, Nicaragua, and Tehuantepec, in relation to their productions and the supply of fine timber for ship-building and all other purposes. With a map and illustrations. London, 1851 : Effingham Wilson.

CORTEZ, HERNANDO.—The dispatches of Hernando Cortez, the Conqueror of Mexico, addressed to the Emperor Charles V, written during the Conquest, and containing a narrative of its events. Translated by George Folsom. New York, 1843: Wiley & Putnam. Pages 99–102, 144, and 360–361 relate to Tehuantepec.

DALE, R.—Notes of an Excursion to the Isthmus of Tehuantepec, in the Republic of Mexico. London, 1851 : Effingham Wilson.

GARAY, DON JOSE DE.—An account of the Isthmus of Tehuantepec in the Republic of Mexico, with proposals for establishing a communication between the Atlantic and Pacific Oceans, based upon the surveys and reports of a Scientific Commission appointed by the projector, Don José de Garay. London, 1846: J. D. Smith & Co.

HERMESDORFF, HERR M. G.—The Isthmus of Tehuantepec, Vol. XXXIII of Journal of Royal Geographical Society, pages 536–554.

KINGSBOROUGH (LORD).—Antiquities of Mexico, 9 vols, London, 1830. Plates 1 to 18 of Vol. IV, Part III, are views of the Antiquities of Tehuantepec. Plates 19 to 45 of Vol. IV, Part III, represent the ruins of Palenque, a little east of the Isthmus. And Plates 78 to 95 of Vol. IV, Part II, are views of the Antiquities of Mitla, west of the Isthmus. These costly and magnificent illustrations must be seen to be appreciated. Plate 82 of the last mentioned list is particularly worthy of mention.

LIOT, W. B.—Panama, Nicaragua, and Tehuantepec ; or, considerations upon the question of communication between the Atlantic and Pacific Oceans. London, 1849 : Simpkin & Marshall.

MURPHY, JOHN MCLEOD.—The Isthmus of Tehuantepec ; its inhabitants and resources. An address delivered before the American Geographical and Statistical Society. In Journal of that Society for 1859.

RAMIREZ, J. F.—Reasons which the Government of Mexico has for not recognizing D. José Garay's privilege to open communication across the Isthmus of Tehuantepec. New York, 1852.

ROBINSON, WM. D.—Memoirs of the Mexican Revolution, including a narrative of the expedition of General Xavier Mina, with some observations on the practicability of opening a communication between the Atlantic and Pacific Oceans through the Mexican Isthmus in the province of Oaxaca and at the Lake of Nicaragua ; and on the future importance of such commerce to the civilized world, and more especially to the United States. Philadelphia, 1820.

STEPHENS, J. L.—Incidents of Travel in Central America, Chiapas, and Yucatan, 2 vols. New York, 1841 : Harper and Brothers. Vol. II, pages 284-358, are devoted to the ruins of Palenque, and contain numerous illustrations of the same.

STEVENS, HENRY.—Historical and Geographical Notes on Tehuantepec, 1453-1869. New York, 1869 : D. Appleton & Co.

STUCKLE, HENRY.—Inter-Oceanic Canals. An essay on the question of location for a Ship Canal across the American continent. New York, 1870 : D. Van Nostrand.

CONGRESSIONAL DOCUMENTS.

SENATE EXEC. DOC. NO. 60, 30TH CONG., 1ST SESS.—Message from the President of the United States, communicating a copy of the treaty with the Mexican Republic of February 2, 1848, and of correspondence in relation thereto, and recommending measures for carrying the same into effect. Page 44 contains the instructions from James Buchanan, Secretary of State, to the United States Commissioner at the City of Mexico, authorizing him to offer Mexico fifteen million dollars for the right of way across Tehuantepec.

SENATE MIS. DOC. NO. 50, 30TH CONG., 2D SESS.—Petition of P. A. Hargous, offering to the consideration of Congress the advantages of a railroad across the Isthmus of Tehuantepec, and praying that Congress, before its final action on the subject, will allow time for establishing the facts therein stated.

SENATE EXEC. DOC. NO. 97, 32D CONG., 1ST SESS.—Message from the President of the United States, in answer to a resolution of the Senate, calling for the correspondence between the governments of the United States and Mexico respecting a right of way across the Isthmus of Tehuantepec.

This relates to the Garay grant, the subsequent assignment of the same to Hargous, and correspondence relative to a proposed convention between the two countries.

SENATE REPORT OF COM. NO. 355, 32D CONG., 1ST SESS.—Report of the Committee on Foreign Relations on the above-mentioned message of the President concerning the Garay grant and assignment to Hargous.

SENATE EXEC. DOC. NO. 72, 35TH CONG., 1ST SESS.—Message of the President of the United States, in answer to a resolution requesting him "to inform the Senate whether any efforts have been made or authorized by the Executive Department or any officer thereof, to induce the government of Mexico to annul or impair the grant of February 5, 1853, for the construction of a plank road and railroad across the Isthmus of Tehuantepec, as recognized in the treaty published at Washington on the 30th of June, 1854, and to obtain a new grant of the same or like character for other parties ; and if so, that he communicate the names of those parties, together with the terms, conditions, and considerations of the grant, and all correspondence connected therewith."

This document relates to the Garay transfer and the Sloo grant.

SENATE EXEC. DOC. No. 25, 39TH CONG., 2D SESS.—Message of the President of the United States, communicating, in compliance with a resolution of the Senate of the 6th of February, 1867, correspondence on the subject of grants to American citi-. zens for railroad and telegraph lines across the territory of Mexico.

This document relates to the La Sere grant for railway, etc., across Tehuantepec.

SENATE EXEC. DOC. No. 6, 42D CONG., 2D SESS.—Report of Survey by Capt. Shufeldt. (See Surveys.)

www.ingramcontent.com/pod-product-compliance
Lightning Source LLC
Chambersburg PA
CBHW032248080426
42735CB00008B/1051